都市里的村庄
Village within a City

[德]里伯斯金建筑事务所等 | 编

司炳月 高松 姜博文 | 译

大连理工大学出版社

004　我们不需要更多的建筑了 _ Richard Ingersoll

韩国建筑的美学——禁欲与放纵的较量

008　禁欲与放纵的较量 _ Mannyoung Chung

018　河阳无鹤路教堂 _ Seung, H-Sang
　　　Between the obsession with temperance and the freedom of redemption _ Hyon-Sob Kim

038　巨济岛宾馆 _ BCHO Partners
　　　Guest House Geojedo; a window converse with the ground _ Kim Young-cheol

058　坡州庭院 _ IDMM Architects
　　　Paju Gardenus; from form to behavior _ Kang Howon

相聚

078　相聚 _ Herbert Wright

086　卡索拉山的"高清"医院 _ EDDEA

104　安德马特音乐厅 _ Studio Seilern Architects

116　克雷斯波洛图书馆 _ Steimle Architekten

132　阿尔科塔农民起义100周年纪念馆
　　　_ [eCV] estudio Claudio Vekstein_Opera Publica

148　MO现代美术馆 _ Studio Libeskind

都市里的村庄

164　都市里的村庄：城市的空间与规模 _ Gihan Karunaratne

172　深圳万科云城B2+A4区 _ FCHA

186　深业上城LOFT _ Urbanus

202　绿洲排楼 _ Serie Architects

216　海军部村庄 _ WOHA

230　建筑师索引

C3 Village within a City

004   We don't need any more buildings_Richard Ingersoll

Abstinence vs. Excess in Korean Architecture

008   Abstinence vs. Excess_Mannyoung Chung

018   Hayang Muhakro Church_Seung, H-Sang
       Between the obsession with temperance and the freedom of redemption_Hyon-Sob Kim

038   Guest House Geojedo_BCHO Partners
       Guest House Geojedo; a window converse with the ground_Kim Young-cheol

058   Paju Gardenus_IDMM Architects
       Paju Gardenus; from form to behavior_Kang Howon

Come Together

078   Come Together_Herbert Wright

086   High Resolution Hospital Center in Cazorla_EDDEA

104   Andermatt Concert Hall_Studio Seilern Architects

116   Kressbronn Library_Steimle Architekten

132   Memorial Space and Monument to the 100th Anniversary of the Alcorta Farmers Revolt
       _[eCV] estudio Claudio Vekstein_Opera Publica

148   MO Modern Art Museum_Studio Libeskind

Village within a City

164   Village within a City: Civic Spaces, Civic Dimensions_Gihan Karunaratne

172   B2+A4/Cloud City of Shenzhen Vanke_FCHA

186   Shum Yip Upperhills Loft_Urbanus

202   Oasis Terraces_Serie Architects

216   Kampung Admiralty_WOHA

230   Index

我们不需要更多的建筑了
We don't need any more buildings

Richard Ingersoll

　　自第二次世界大战结束至今，人类建造的建筑物在数量上已超过此前7000多年建造量的总和，虽然不可能去证明，但这显然就是事实。1927年世界人口总量为20亿，如今已达到70亿，人口的急剧增多带来了前所未有的建筑热潮。随后，农村地区也迅速减少，一半以上的人口聚居于城市。所以看到以上这些数据，"我们不需要更多的建筑了"——我的这种主张听起来很可能有些荒唐，并且我这样的表述肯定会被解读成一种对建筑业的攻击。但请不要误会我的意思：我热爱建筑，我永远尊重建筑同业者，坦率地说，叱骂这个行业对我个人没有好处，毕竟我也置身于这个行业，也需要依靠这个行业来谋生。但是，无论在哪里，人口稠密的都市也好，废弃的乡村也好，所到之处我总能看到有些建筑空荡荡地杵在那里。这让我反思为什么还要建造更多的建筑物。我还注意到建筑物中很大一部分的结构没有得到充分的使用，它们常常彻夜灯火通明，显得非常浪费。我问自己为什么这么多当代城市被规划设计为单一用途？对建筑空间需求迫切的人们和被闲置浪费的建筑空间之间的不协调，促使我得出以下结论：我们不需要更多的建筑，我们需要的是对建筑进行更合理的分配。

　　但是，我的质疑很可能会遭到一些人简单粗暴的反驳，他们会说："这就是经济问题了，蠢货"，因为任何一个成功的城市，至少有三分之一的财富都来自建筑。在自由经济盛行的环境下，开发商和他们的设计师受利益的驱使，在生产创造过程中，必然采取一系列有害的手段，譬如，拆毁旧建筑、攫取资源、购买进口材料建设房屋。这种不破不立的习惯源自于人们疯狂依恋于碳氢化合物的工业历史，并一度被视为一种"进步"。可是，随着人们对气候变化的认识越发深刻，这种习惯必须重新被归为"破坏"。任何的建设性行

Although it is impossible to prove, it seems obvious that in the period since the end of World War II more buildings have been produced than during all of the 7 millennia of human construction activity that preceded it. This unprecedented building boom has accompanied an exponential population increase, from 2 billion in 1927 to 7.7 billion today, along with a radical depletion of rural areas, resulting in well-over 50% of humanity residing in urban situations. So considering these numbers it might sound absurd for me to pronounce "we don't need any more buildings", and such a statement will certainly be interpreted as an assault against architecture. Please don't get me wrong: I love architecture and have an undying respect for the profession, and frankly it would not be in my own interest to abuse the field, seeing that I make my living from mediating its output. Everywhere I look, however, I see empty buildings, both in dense urban environments and in abandoned countrysides, and it makes me wonder why more have to be built. I also notice an abundance of underused structures, often with their lights burning throughout the night, a sign of extravagant waste, and ask myself why so much of the contemporary city has been programmed for single uses? The mismatch of people in need and empty space prompts me to conclude that there are enough buildings already, we just need better allocation.

A simple rebuttal to my queries, however, might be: "It's the economy, stupid," since at least a third of any successful city's well-being turns around construction. In the liberal economy that prevails, the creative energy of developers and their designers gets inevitably channeled by profit-driven motives toward the deleterious cocktail of demolition, extraction, and construction with imported materials. This habit of creative destruction, inherited from the carbonholic industrial past, was once regarded

为，无论多么"可持续"，都会加剧即将到来的地球生物灭绝的困境，而不是平息，只不过是程度不同罢了。

同时，隐藏在许许多多当代建筑作品背后的虚荣与贪婪，造就了历史上的一些造价极高、规模巨大的建筑，更别说那些微不足道的建筑了，而全球变暖的问题也并没有得到缓解。即使是那些获得LEED铂金认证的环保建筑，虽用心良苦，但终究也是把这种殊荣当成了一种营销手段，一边把自己"漂绿"树立着虚假的环保形象，一边服务于那些乐此不疲地压榨人力、耗费不可再生能源的投机企业。欧洲最高的建筑物，即伦敦的The Shard大楼，在开放十年后仍然很明显地空置在那里，这就是一个现成的案例，表明这个项目的优先事项放错了地方。这座大厦在设计上秉持了一些可持续发展的理念，例如，增加城市密度、采用双层玻璃减少能源需求……说是这么说，可实际上这个庞然大物赫然耸立在泰晤士河的河边，显得十分多余，当初吆喝着要把它设计成伦敦的埃菲尔铁塔纯粹是营销噱头而已。在一个公正合理的社会中，任何闲置超过两年的可居住闲余建筑都应课以重税，并最终被征收，物尽其用。既然百万富翁们不买账，那就把它分给人民吧！

巴塞罗那真就颁布了这样的政策，这表明我们的可持续发展任务仍然面临着种种矛盾。开发商们疯狂地搞起了投机建设，一个个争先恐后地建造更多的旅游度假区，为了遏制这种猖獗的行为，左派市长艾达·科洛下令对数千个空置住宅实行新的税收政策，这些空置住宅主要由银行和控股公司所有。如果业主同意依据社会租赁条件出租这些地方，使那些被价格飞涨的住房市场拒之门外的人们得以安置，则政府

as "progress," but since the awareness of climate change must now be reclassified as damage. Any act of construction, no matter how "sustainable", contributes to inflating rather than calming the planetary dilemma of impending biological extinction. It is a matter of degrees. Meanwhile, the combination of vanity and greed that lurks behind so much of contemporary architectural production has unmolded some of the largest and most costly structures in history, not to say the most frivolous, while failing to seriously cool down Global Warming. Even the well-intentioned projects that obtain LEED platinum ratings, use such distinction as a marketing device, a form of green-washing, while remaining in the service of speculative enterprises that refuse to relent in their exploitation of people and non-renewable resources. That the so-called Shard in London, Europe's tallest structure, remains conspicuously empty a decade after opening, provides me a ready example of a monument to misplaced priorities. Designed with sustainable objectives, such as increasing urban density, and reducing energy needs through double glazing, it looms over the Thames as a redundant icon, a pure gimmick of marketing meant to give London the equivalent of the Eiffel Tower. In an equitable society, any habitable empty space that remains vacant for over two years should be heavily taxed and eventually expropriated for social uses. Seeing as the millionaires are not buying into it, let's give it to the people!

Such a policy was indeed enacted in Barcelona and remains indicative of the contradictions we face in the mandate for sustainability. The left-leaning mayor Ada Colau ordered a new tax policy on thousands of empty dwellings, mostly owned by banks and holding companies, hoping to curtail the rampant speculation that developers practice in the race to create more tourist structures.

将取消对这些业主的罚款。然而,加泰罗尼亚地区的最高法院最近推翻了这位市长的政策,该法院坚称该市无权对私人财产施加此类限制。

我对废弃建筑的兴趣发生在50年前,当时我是一名志愿者,参与修复托斯卡纳区一个几乎被遗弃的城堡小镇。森尼纳曾有300人居住,当时仅剩下10人。其余的人都搬到城市去了,他们在工厂和服务业工作,偶尔开着锃亮的汽车返回他们的村庄,但已不再是那里的居民。我们没用多么高超的技巧,就是维修了瓦屋顶,给它们加了保温层,并在破旧的房屋墙壁上进行了抹灰处理。城堡的修复工作则由专业人员进行,半个世纪后这座城堡还未完工。从那时起,我就一直在寻找一些废弃的建筑物并住进去,有时甚至是不打招呼就直接占据其中。最终,我寻获了一座废弃的农舍,这座200多岁的建筑几十年来一直没有人住过。我发现,比起建造时髦的新建筑,修复旧建筑让它重新投入使用所需要的创造力更多,而且不止我一个人注意到这一点。

最近,我和Dante Bighi基金会在费拉拉市郊一个叫科帕罗的小镇举行了一次研讨会,该镇曾经是繁荣的农业中心,后来又进行了以钢质机械部件为主的工业生产。目前,由于该镇的工农业都在衰败,许多人(特别是年轻人)都已经离开,给市政当局留下了超过20 000m²的废弃建筑。我们的提议是,任何愿意来科帕罗整修这些建筑物并发起一项创意活动的人都可以免交房租,专心为城市的公共事业做贡献。可惜的是,邀请我们的这些政治活动家在最近的选举中失败了,我希望不是因为我们的提议造成的,而这些改造想法也只好继续搁浅了。

The fines would be rescinded if the owners agreed to lease the space according to a social rent to accommodate those who have been excluded from the inflated housing market. Her policy, however, was recently overturned by the Catalan region's Supreme Court, which insisted that the municipality does not have the right to impose such restrictions on private property.

My interest in abandoned buildings began 50 years ago – when I worked as a volunteer on the restoration of nearly abandoned castle town in Tuscany. Of the 300 people that once had lived in Cennina, only 10 remained. The rest had migrated to the cities for work in factories and services, occasionally returning to their village in shiny cars, but no longer residents. Without much skill we fixed and insulated the tile roofs and plastered the walls of the old houses, while professionals worked on the restoration of the castle, which, half a century later, is still unfinished. Ever since then I have always looked for and lived in abandoned buildings, sometimes squatting them. Eventually I acquired an abandoned 200-year-old farmhouse, which had not been lived in for several decades. I am not alone in observing that more creativity is required in fixing things up for reuse than creating buildings ex-novo.

I recently organized a workshop with the Fondazione Dante Bighi in Copparo, a small town outside of Ferrara, once a thriving agricultural center, which later hosted a mix of industrial production, mostly involved with steel mechanical components. At present, both industry and agriculture have declined and many people, especially the young, have left. The municipality currently owns more than 20,000m² of abandoned buildings. We proposed that anyone who would come to Copparo, fix the unused buildings, and initiate a creative activity could do so rent free, contributing only to the services of the city. Alas, the political exponents who invited us lost the recent election, I hope not because of our proposals, and such ideas will have to wait.

在自由经济中，可持续发展的经典原则，譬如，"减量化、再使用和再循环"似乎与盈利背道而驰。虽然一些理论家，如Serge Latouche，鼓吹"退行生长"，但地球上任何一个政治家如果还想继续干，就不可能预想减少发展这种主意。发展仍然是自由主义的圣杯，并意味着成立强大的建设部门。即使是联合国这样致力于遏制气候变化这一崇高使命的机构，也难以避免发展经济的问题，从而推出了"可持续发展"，而不是完全减少或消除增长。因此，我开始钦佩那些回收多余空间并将其再利用的人的做法，这似乎是我们能做的最有创意和最可持续的事情。

法国建筑师Lacaton和Vassal曾致力于改造丑陋的预制装配式房屋，例如，波尔多的Le Grand Parc，按照他们的推算，如果选用一些常规的加固结构作为基础来包裹这些预制房屋的外壳，这些外壳的支架就额外扩出了一层可居住空间。整套作业不仅优化了每个单元的采光、增强了保温性，而且还额外给住户增加了20m²的空间，同时在施工过程中也不用把住户迁出去。

我虽然没去过韩国，但我读过的一些文章上面说，首尔是地球上人口最密集的城市之一，所以很可能还没开始出现那种"空楼"现象。但为什么大家都要挤进首都居住呢？特别是现在这个时代，我们有了数字技术，几乎可以在任何地方做任何事情（除了建设工作）。如果我们可以回到村庄，把旧建筑修好，增建一些基础设施，再参与一些农业福祉活动，我们何必还要建更多的新建筑呢？

In a liberal economy, the classic principals of sustainability, "reduce, reuse, and recycle" seem antithetical to making a profit. While some theorists, such as Serge Latouche, preach "degrowth", no politician on the face of the earth can survive by predicting the idea of less development. Development remains the sacred grail of liberalism, and implies a strong construction sector. Even the United Nations' agencies devoted to the virtuous mission of arresting climate change have difficulty avoiding the issue of growth and thus promote "sustainable development", rather than reducing or eliminating it altogether. Thus my admiration goes to those who occupy and reuse, which seems to be the most creative and sustainable thing we can do.

The French architects Lacaton and Vassal have made a career of retrofitting ugly prefab housing estates, such as Le Grand Parc in Bordeaux, reasoning that it is more costly and environmentally damaging to demolish such buildings than to use their generally solid structural base for wrapping the exterior with a scaffold that provides an extra layer of habitable space. This process has improved the daylighting and thermal performance of the units, while giving the occupants 20m² more room, and it did not require evictions.

I have never been to Korea, but I have read that Seoul is one of the densest cities on the planet, and perhaps the "empty-building syndrome" has not yet arrived there. But why does everyone have to live in the capital? Especially now that we have digital technologies that make it possible to do almost anything (except construction) anywhere. If we returned to the villages, fixed up the old buildings, added a bit of infrastructure and participated in their agricultural well-being, would we need any more new buildings?

韩国建筑的美学——禁欲与放纵的较量

Abstinence
in Korean

本章将介绍三位韩国建筑师的最新作品。

建筑师承孝相的"河阳无鹤路教堂",他形容它是"一座只留下最基本的元素,让我们在此自省、与上帝沟通的教堂";建筑师赵秉洙的巨济岛宾馆,他说"这是一座饱含了他对大地的信仰、暗含着他可持续性设计理念的建筑";建筑师郭熙秀的"坡州庭院"对院子、混凝土凉亭和平床(一种低矮的木凳)这些元素重新进行了诠释,他表示:"这座建筑可以同时满足想要将其空间用于公共用途的人和想要用于营利活动的人,为他们彼此提供了方便。"三件作品都反映了韩国当代建筑的多样性,因为它们忠实地投射出建筑师的标志性特征和设计风格。

本书特邀四位建筑评论家对以上三座韩国建筑进行了考察,并对它们的设计者进行了专访。经过共

This chapter introduces the most recent works of three Korean architects.
"Hayang Muhakro Church" – which, in the words of its architect Seung, H-Sang, "was built as a church that allowed us to reflect on ourselves and communicate with God by leaving only the most essential things"; "Guest House Geojedo" – by Byoungsoo Cho, who said that it contains his belief in the land and his implicit story about sustainable architecture; and "Paju Gardenus" – which reinterprets the elements of courtyard, concrete pavilion, and low wooden bench (Pyeongsang), and, as its architect Heesoo Kwak said, "serves as a mutually convenient place for those who want to use the architectural space for public use, and those who want to pursue profit-making activities." All three works reflect the diverse spectrum of Korean contemporary architecture, in that they faithfully project their architects' defining characteristics and styles.
The four critics who responded to the book explored these three buildings across Korea and then

河阳无鹤路教堂_Hayang Muhakro Church / Seung, H-Sang
巨济岛宾馆_Guest House Geojedo / BCHO Partners
坡州庭院_Paju Gardenus / IDMM Architects

禁欲与放纵的较量_Abstinence vs. Excess / Mannyoung Chung

vs. Excess Architecture

同的考察和采访后,他们共同承担了这次专题,交换了彼此的意见,并把各种观点进行了整合。

郑万荣扩展了三位建筑师的时间视域,为整体大环境设定了场景。金玄燮对建筑师推想的"贫无之美"的由来进行了详细交代之后,评价了河阳无鹤路教堂的几何形式的抽象性及其简单清晰的空间构图。金永澈对巨济岛宾馆的设计给出了解释,他说设计师赵秉洙的设计初衷源于他一直以来对"大地"的关注,但后来他令这座建筑重生为连接天空与自然的场所。他写道:"当我们坐在那里时,世界如建筑师说的那样被'璀璨的星空'所填满,它没有具体的形状,只有空间的存在,成为大地、天空和自然之间的纽带。"最后,姜浩元将坡州庭院的特征定义为"书法美学",并强调了建筑师郭熙秀的关注点开始从形式转向行为。

interviewed the architects. Having gone through these together, they were able to share the context of this special feature, and to exchange and organize many diverse opinions.
Mannyoung Chung extended the temporal horizon of the three architects and set the scene for the overall context. Hyon-Sob Kim looked at the abstraction of the geometric form of the Hayang Muhakro Church and its simple and clear spatial composition, in the context of expanding the architect's reasoning of the "Beauty of Poverty". Kim Young-cheol explained that Guest House Geojedo began with Byoungsoo Cho's long interest in "Earth", but that it was reborn as a place connecting sky and nature, writing: "When we take a seat there, the world is filled with 'a night sky with stars that shine brilliantly', and a place without form, where only space exists, becomes a connector between the earth, the sky and nature." Finally, Kang Howon defines the characteristics of Paju Gardenus as the aesthetics of Il-pil-seo (calligraphy), and insists that Heesoo Kwak's interest is shifting from form to behavior.

禁欲与放纵的较量
Abstinence vs. Excess

Mannyoung Chung

　　本书对来自韩国的三位当代建筑师承孝相、赵秉洙、郭熙秀的作品做了一次专访。由于他们每个人都秉持了不同的设计理念,因此本书的主题会与以往有些不同,但同时也引起了读者强烈的兴趣。现如今,如果将不同的个性混杂在一起已成为一种时代特征,那么这次的专访合集在行文构思上则要减少这种混杂感,将这些风格不同的设计作品以某个线索串联起来,找出一个交集,这个线索就是设计者对自己以往信念的坚守。

　　承孝相的河阳无鹤路教堂体现了他所坚持的"贫无之美",从他1992年设计守拙堂以来,这一直是他设计的指导理念,尽管这个新作与他1994年设计的、位于唐津市的九德天主教堂十分相似。赵秉洙的巨济岛宾馆可以归为"大地屋宇"系列之一,这个主题自他毕业以来就一直在他的作品中延续。其他代表性作品包括数谷里大地屋宇(2009年)、数谷里混凝土盒子屋(2004年)、LeeOisoo画廊(2009年)、济州岛冥想阁(2010年)和数谷里斜屋顶住宅(2014年)。郭熙秀的坡州庭院十分类似于他之前的一些混凝土建筑系列的设计,例如,清潭洞Tethys(2007年)和机张Waveon咖啡馆(2016年)。

　　出人意料的是,这三位建筑师都是在37岁的时候成立了自己的设计工作室(承孝相在1989年,赵秉洙在1994年,郭熙秀在2003年)。他们为了赢得业界认同,各自拼命奋斗,所以设计室成立不久就声名鹊起,各自的设计风格都在业界独树一帜。他们当时提出的设计思想和美学理念在本书展示的三部作品中都得到了充分体现,因此,让我们进一步了解一下随着时间的发展,他们的建筑语言是如何发展的,目前应该从什么方面评价。

　　承孝相在早期的设计生涯中,要克服的最大难关就是"摆脱老师金寿根的设计风格"。如果一个人在别人的影响下成长,却没有能力从别人的麾下走出来,那这个人就永远只是模仿前人的庸才。美国文学理论家哈罗德·布鲁姆将这种独特的关系概括为"影响的焦虑"。为了克服这种影响,新一代的艺术家们拼命努力,正是这种动力将他们的创造力发挥到了极致。他的前辈资历越高,对他产生

The special feature in this book on the work of three contemporary Korean architects, Seung, H-Sang, Byoungsoo Cho, and Heesoo Kwak, each of which takes a distinctive direction, gives a bit unusual impression but also arouses intense curiosity. If the mixture of individualities is a characteristic of the contemporary world, this feature is reducing its arrangement to a similar subset. The clue to crossing this asymmetric composition lies in the fact that three architects are closely bound with their own past respectively.
The "Hayang Mahakro Church" of Seung, H-Sang embodies the "Beauty of Poverty" has been the architect's guiding philosophy since Sujoldang (1992), albeit it is strongly reminiscent of the Dolmaru Catholic Church, Dangjin (1994). Byoungsoo Cho's "Guest House Geojedo" can be categorized as one of a series of "Earth Houses", a theme continued in his work since his graduation project. Other representative works include: Sugokri Earth House (2009), Sugokri Concrete Box House (2004), Lee Oisoo Gallery (2009), Jedong Ranch in Jeju: Meditation House (2010), and Sugokri Tilt Roof House (2014). Heesoo Kwak's "Paju Gardenus" also closely resembles this architect's concrete series from Tethys (2007) and Gijang Waveon (2016).
The three architects unexpectedly opened their offices at a similar age of about 37 years old (Seung, H-Sang in 1989, Byoungsoo Cho in 1994, and Heesoo Kwak in 2003). Shortly thereafter, through a fierce struggle for recognition, they created strong and distinctive identities for their architecture. The thoughts and aesthetics of this process are fully projected in the three works published in this book, which in due course will broaden the horizon of discussion as to how their architectural language has expanded, and what can be evaluated at this point.
The most important issue in the early career of Seung, H-Sang was "getting away from Kim Swoo-geun". If anyone grows under the influence of someone else, but fails to demonstrate his or her ability to overcome it, he or she will always be a lite of predecessor. Harold Bloom, an American literary theorist, has summarized this unique relationship as "anxiety of influence". The struggle to overcome this influence is the driving force which brings the junior artist's creativity to its peak. The greater the seniority, the greater the influence, and the greater the achievement to overcome. For Seung, who played a core role within the architecture of Kim Swoo-geun, self-reliance meant that he denied himself as well as

九德天主教堂,承孝相,唐津,
1994年
Dolmaru Catholic Church by
Seung, H-Sang, Dangjin, 1994

的影响就越大,而他克服这种影响之后取得的成就也会越高。对于已经是金寿根建筑体系中核心人物的承孝相来说,独立意味着他否定了先前的自我,也拒绝了对老师的复制。

承孝相经过努力,找到了两个途径发挥自己的潜力,帮助自己从金寿根的风格中独立。

首先,他否定了金寿根的建筑主张。事实上,他用简单和严谨的几何构图表达了符合自己气质的建筑语言。但另一方面,他本来也可以沿用金寿根主张的自由几何构图,那也能让他的建筑很有影响力。可是,重要的不是做他很擅长但属于别人的东西,而是哪怕经过一番艰苦奋斗也要实现自我超越。

此外,承孝相使他的建筑和他的建筑理论保持了一致性。他在1996年发表的《贫无之美》给韩国建筑界证明了,基于理论指导下的建筑实践是能够实现的,或者说具有实践性的建筑理论是存在的。在此之前,也有许多人尝试用一些理论来陈述建筑的合理性。一些建筑师试图向人们解释他们的建筑和自然或传统之间的联系,金寿根也在其中。他通过各种调研,发明了"极致空间""子宫空间"和"否定主义"之类的词语,设法把它们强加到自己的建筑论述之中。然而,当时的建筑和相应的语言彼此格格不入,两者几乎从未统一过。承孝相的"贫无之美"对建筑史的意义在于,他成功地把他的建筑语言凝聚到了建筑中,那些已完成的作品就是证明。他秉持的"贫无"哲学也是在回应20世纪80年代后现代主义折射出来的浮华与虚荣,以此为契机让人开始重新反思建筑内涵。

河阳无鹤路教堂(第18页)占地面积8.4m×8.4m,内外都采用了砖饰面。整座建筑高耸封闭,加之周围又都是些歪歪扭扭的单层住宅和辅助性建筑,使自身更加引人注目。不过,它并没有让人产生压迫感,它周围有足够开阔的空地来缓解整个区域的视觉压力。整个外墙除了一个入口,再没有设其他门窗,但这没有使建筑显得呆板。教堂如同一个小方盒被半隐在大的方盒中,外面环绕着围墙,

Kim Swoo-geun.
Seung secured two clues to identify his own capabilities in the struggle to overcome Kim.
Firstly, Seung denied Kim's formative architectural language. The fact that his architectural language resulted in a simple and strict geometry reflects his own temperament, but on the other hand, Kim's reaction to the free form would have played a major role as well. It's not just what he can do well, but it's a new competency that he gained through a bitter struggle for self-overcoming.
Secondly, Seung assigns his architecture a theoretical identity. It was not until Seung's Beauty of Poverty (1996), that Korean architecture confirmed the realization of theoretical practice or practical theory. Prior to that, there were many attempts to justify architecture with theoretical statements. Like many architects who tried to explain their architecture in connection with nature or tradition, Kim Swoo-geun tried to impose his own words on architectural discourse through various investigations such as "ultimate space", "uterine space", and "negativism". However, the architecture and the corresponding language at this time were an amalgam of two distinct fields which were barely ever unified. The historical significance of Seung's "Beauty of Poverty" lies in that he was able to prove a cohesion between his words and built architecture: "the words that correspond to the architecture" and "the architecture integrated with the words". The "Beauty of Poverty", which was also a reaction to the vanity of the postmodernism of the 1980s, presented an opportunity to reflect on architecture.
"Hayang Muhakro Church (p.18)" is a volume of 8.4m × 8.4m in size finished with bricks inside and out. It is a high, closed block, so it is more noticeable than the crooked surrounding single-story houses and ancillary buildings; but it is not overwhelming, and it has enough open area surrounding it to loosen the context. There is no opening in the outer wall except for the entrance, but neither is it bland. A box (the church building) is nestled inside another square box surrounded by an outer wall, and a small, long box (the open-air prayer room) sits on top. When viewed from the east, the facade appears flat. But the southern part is about 1.5m higher than the northern, which resembles the typical shoulder-shrugging posture of a Korean dancer.

坐落在教堂顶部的长方体盒状结构是露天的祷告室。从东侧看，建筑的立面很平坦。南侧结构要高出北侧1.5m，象征着朝鲜舞中典型的耸肩姿势。

在外层盒子和内层盒子之间的空隙处能直接看到天空，正面就是入口处的走廊，侧面是沿着对角线上升的室外楼梯。外边的盒子刚好把内部的入口遮住，因此在人们穿过走廊进到教堂的过程中，会感觉到细节的变化。例如，横贯整个空间的空隙和密实的围墙之间产生的律动，竖直方向狭长逼仄的墙面和水平方向的构图视线形成的对比，还有将墙的平面性与建筑物的空间性相互交错产生的层次感。整个"盒子"的内侧构图变幻莫测，引人入胜，给外侧严密封闭的观感来了一个大逆转。

然而，这么解读他设计的教堂的空间，承孝相似乎不那么满意，直言："（如果是这样）那我还是没有跳出窠臼……"他在项目总结中表明了他对教堂设计的立足点："这座教堂不是为了上帝的留居而建，而是为了那些听到上帝召唤、被尘世的种种欲望放逐到此的人们而存在。既然如此，这座教堂的设计不是应该回归到最简单的形式吗？这样才能反映教堂的本质不是吗？"这位建筑师提出问题后又自己给出了答案："无鹤路教堂的空间结构简单清晰，人们来到这里寻求自由和真理，教堂为这些探求者们而存在。"建筑师只是建造了教堂的大楼，并没有给它设计附加设施，为的就是把教堂的几何空间简化到极致。而且，整个空间一目了然、浑然一体。承孝相也用同样的观点在杂志《石枕》（2019年）中对保罗·瓦莱丽关于罗马万神殿的新书《冥想》进行了总结："最神秘的品质莫过于清晰明澈。"在这种情况下，前面说的靠着突出视觉层次感来表现美学魅力的做法不过是一种半吊子的禁欲。

"将自己逐离出世俗的欲望"是承孝相的本体论哲学立场，这句话表达了承孝相对宗教的思考：人通过救赎获得自由。他的建筑也在表达同样的思想。他让自己专注于最本质的东西，追求城市共享，摒弃形式上的表现，远离责难与对抗。他说，贫无之美已经在扩大。也许光是唯一可以介入这个空间的媒介，人们在此得到精神的升华，专注地感知整体的存在，而不是追求视觉变化或美学情趣这些形式上的东西。只有那些被放逐到黑暗和寂静中的人才能体会到光影之间不可思议的微妙境界，但这种超然物外的世界似

The gap between the enclosing box and the enclosed box is open to the sky. The entrance passage is on the front and the outdoor stairway rises diagonally on the side. The upper box covers the entrance door. As a result, when walking along the entrance passages, small changes are perceived, such as a rhythm of the pierced void and the solid volume, the contrast of the tightly squeezed vertical wall and the horizontal sight-line, and the juxtaposition of the two-dimensional wall and the three-dimensional volume. It's an unpredictable, attractive reversal outside of a tightly closed mass.

However, Seung seems to be displeased at these traits to be referred to blurting out: "I yet haven't thrown it away …" Work summary shows his foothold for the church design: "The church is not the house of God, but the house of those who are called, those who have been exiled from the desires of the world. So shouldn't the church be the simplest form – a form of the essence?" The architect answers his own question: "Muhakro Church has a clear and simple space. It is the house of those who search for freedom and truth." The architect built only the church building, without the additional facilities, because he was trying to create a space that was reduced to the simplest geometry. Moreover, the whole was grasped at a glance, and the whole was united. In the same vein, the architect sums up Paul Valerie's words in his new book Meditation (Dolbegae, 2019) about the Roman Pantheon: "There is nothing more mysterious than clarity." In this context, aesthetic appeal, which depends on the ability to discriminate, is merely an immature lack of abstinence.

Seung, H-Sang's ontological position, to be "the one who has exiled himself from the desires of the world" is another expression of religious reflection, and freedom through redemption. His architecture works the same way. To share the city, to erase the formal expression, to be free from reprobation and confrontation, Seung restrains himself to the basic and essential. Seung says that the "Beauty of Poverty" has expanded. Perhaps light is the only medium that can intervene in the space where mental thought is deepened, the space for concentrating on the whole presence, not on visual change or aesthetic appeal. Only those exiled into darkness and silence can appreciate the incomprehensibility of light. But this transcendental world seems too far away.

河阳无鹤路教堂北侧一景
Hayang Muhakro Church, seen from the north

从采光天窗照进来的光线照亮了布道台
Light from the skylight illuminates the sermon

乎太遥不可及了。

赵秉洙出版了他的作品集《大地屋宇》(空间丛书，2017年)，里面每个建筑空间都深入到地面以下。这种深入土地内部的设计富有诗意，成为作品中最有影响力的元素，也巩固了他的设计师地位。他在书中说到他看过朋友母亲的墓，是一个"由红泥做成的、截面为方形的空间"(第14页)，唤起他一种本能上的感受，刻骨铭心。他沿着这种原生的本能体验往下挖掘，"生成"了一种向地下延伸的建筑结构，这种经验也成就了他独特的设计手法。赵秉洙关注他想强调的极小部分元素，其余的全部"消除"。

但是要关注哪些、消除哪些？他说："大地的故事是我们在大地上凝望天空时得到的故事，微风吹拂过大地时讲了它的故事，云朵从一个地方飘过时也留给大地它的故事，还有雨……"(第8页) 这是一种现象学，因为直接通过感觉获得的体验要先于建立在理性或视觉上的理解。在这里，建筑成了扩大、加深和表达感官世界的工具。"既然是贴近大地的建筑，就一定要和大地本身一样简单纯朴，融为它的一部分。要想达到这种极简的效果，必须合理地安排空间布局和建筑材料，有时候甚至与建筑本身的结构和施工方法有关。"(第22页) 让房屋贴近土地的设计很简单，这意味着形态的运动、空间的变化，以及构造体系的倒退。相反，触觉效果被放在了第一位，以有效地增强感官体验。整座建筑被做成不规则的形状，这使得建筑空间无法被分割成一个个小单元，这样也就无法对它施加限制了。世间万物处于永恒的变化之中，而这种建筑的"无形"性就是要把这种变化带给人的原始感受给表现出来，用光、风、声、水这些无形的自然物去"描绘"这种感觉，任何形式上的安排都会限制这种原始感受的充分表达，通过自然物质来形成一种与大地贴合的原生体验正是这种建筑的魅力所在，给予的体验越是强烈，它的魅力就越能得到显现。

巨济岛宾馆(第38页)很典型地呈现了赵秉洙"让建筑贴近土地"的设计理念。来到附近的游客可以把建筑当成一个水平的看台，视线可以丝毫不被遮挡地望向大海。建筑失去三维立体感，只在地面上勾勒了二维的边界。建筑空间层次的动态感被消除，它的存在可由一些线条来确认，如"画"在地上的边界之间存在的豁口、向远处延伸成斜坡的墙垣、把屋顶分成台阶的屋顶线，从地表向

Byoungsoo Cho published a series of architectural works, each dug into the ground, in his book House in the Earth, House towards the Earth (Space Books, 2017). His technique for designing within the land is the most powerful poetic device that has solidified his identity. In the book, he says that seeing "a square space made of cross sections of red mud" (p.14), the grave of his friend's mother, has provoked a primitive experience that imprinted on him intensely. This primitive experience developed into an architecture which dug into the ground, becoming a unique modus operandi. Cho focuses on the minimum he wants to emphasize and erases everything else.

But what to focus on, and what to eliminate? He says: "The story of the earth is the story of the sky that we gaze up at from the earth, it is the story of the breeze that plays over the earth, the story of the cloud that passes over terrain leaving the earth in its wake, the story of rain, … " (p.8) It is closely related to phenomenology in that direct experience through the senses precedes rational or visual understanding. Architecture becomes a device to expand, deepen and frame the sensory dimension. "Architecture towards the earth must also be simple. It must be as simple as the earth so that it can become a part of the earth. This simplicity must be defined in relation to the structure of the space, the material used, and sometimes even in relation to the structure itself or the construction methods employed." (p.22) The fact that architecture towards the earth is simple means that the movements of form, variations in space, and the tectonic system move back. Instead, tactile properties come forward to reinforce effectively sensory experience. This is because amorphous properties, which are not possible to segment into units or set limits for, respond to the primordial sensation of the ever-changing body, depending on how it encounters light, wind, sound, and water. Everything that is organized in form only constrains this primitive sensation. The appeal of architecture towards the earth depends on the extent of the intensity to ascertain the primordial sensation through materiality.

"Guest House Geojedo (p.38)" shows typical characteristics of Cho's architecture towards the earth. The house is located between the mountain and the sea. For visitors approaching the site, the building becomes a horizontal podium that assists in reaching out to the sea without obscuring the view. Architecture loses volume and remains a boundary, with lines drawn on the ground. The movement of the architecture is eliminated and its existence is confirmed through the

数谷里斜屋顶住宅, 2014年
Sugokri Tilt Roof House, 2014

巨济岛宾馆, 2018年
Guest House Geojedo, 2018

下彼此分隔的外墙线。建筑被"最小化"之后,天空的存在感就会格外地强烈,人们可以眺望远处的大海与蓝天,听见风和海浪的絮语,神驰其中,这便是在感受自然、呼唤自然。想要体味诗意,就需要静静聆听,驻足等待,放空思想,只是匆匆地经过是捕捉不到什么诗情画意的。你靠在咖啡馆的墙边,望着土地、房屋和远处的大海,形成了景中有景、画中有画的纹心结构,把整个房屋的方方面面都压缩在一幅画中。

当然,赵秉洙没有完全走这个设计路线,深入的地下建筑只是这座房屋的一个设计要素,用来反映他极致的自我意识,面对各种现实因素,这种设计自然是不能一一满足的。可是,我们不能忽视这种房屋给了他实验上的动力,使他能用各种性质的材料以不同的方法进行尝试。尤其是随着设计规模的扩大,他设计选材的焦点也从玻璃转移到了金属上,同样也是简化建筑外形,着重通过对材料的把握来表现不同的表皮效果。

从赵秉洙的一系列设计作品可以看出,他在对各种材料进行实验,验证它们对建筑表皮产生的效果。最初设计的双子塔(2010年)采用的是曲面幕墙的表皮设计;在Kiswire中心(2014年)的表皮设计中,开始着重对金属材料的利用;在F1963综合建筑(2016年)的翻修设计中更是采用了金属板网壳作为表皮材料。再往后,在朴泰俊纪念馆(2017年)中采用了和砖体大小相符的耐空气腐蚀钢材,以及组合式压型铝曲面板;画家朴栖甫住所兼展馆(2017年)使用了多孔式厚铝板表皮;天安现代汽车全球运行中心(2018年)则是通过调整金属网切件的角度,实现了表皮疏密渐变的质感。一般来说,通过改变金属网架的孔隙尺寸和角度,可以使建筑外立面持续地随着光影的变化而呈现出不同的观感,同时也能控制亮度和视野。

一边是贴近土地的"大地屋宇"系列的建筑,一边是对金属做各种变形处理、在表皮呈现效果上进行实验的系列建筑。在时间轴上,这两个系列是并行产生的。虽然表面上它们看上去差别甚远,但有一点值得我们注意,那就是它们的着力点都在于突出材料的物理特性和外观质地,而不是拼命地表现建筑形式和空间。这其中的原因也跟设计师过去的经历有关。换句话说,"建筑材料原本

gap between the boundary drawn on the ground, the wall extending to the slope, the roof line dividing the step, and the exterior wall partitioned separately from the ground. Architecture does not reveal itself, but evokes something. Architecture is minimized to make the sky intense, to look at the sea and clouds far away, to listen to the wind and the sound of the waves, that is, to beckon nature. There's not much to capture while rapidly passing through. The poetry takes place only once we listen quietly, wait, stay and empty our minds. Leaning against the wall of the café and looking at the land, the house and the sea beyond, it is a kind of mise en abyme that compresses all aspects of the house.
Of course, Cho is not taking this line entirely. The architecture which digs into the earth is only one side of the property, reflecting an extreme self-consciousness. It is not enough to confront miscellaneous situations of reality. However, one must not overlook that this attribute is at the root of his architecture. The properties are the driving force for experimenting with the different material attributes in different ways. In particular, as design scales increase, his attention quickly shifts to glass and metal. Here too, the form is relatively simple, and attention is focused on surfaces whose effects vary depending on the way the material is handled.
Beginning with the curved curtain wall of the Twin Tower (2010), through the Kisswire Center (2014) which was an opportunity to use metal in earnest, the expanded metal used in the renovation of F1963 (2016), atmospheric corrosion resting steel to fit the scale of bricks, and an assembly type of curved extruded aluminum panels in Park Tae-Joon Memorial Hall (2017), perforated aluminum plates in Park Seo-Bo's house and exhibition hall (2018), and gradual change in surface density by adjusting the angle of the expanded metal cutting part for Hyundai Motor Global Running Center in Cheonan (2018) – these cases show a pattern verifying various surface effects through material experiments. In general, by varying the perforation size and angle of the expanded metal, the exterior facade varies every time with change in light and shadow changes, while simultaneously controlling the lightness and view.
In chronology, the series of buildings towards the earth, and a series of works that examine the surface effects by applying various deformations to the metal, occurred at the same time. They may seem quite different approaches, but it is worth noting that both prioritize the physical properties and surfaces of materials, rather than actively

F1963综合建筑,2016年
F1963, 2016

Kiswire博物馆及培训中心,2014年
Kiswire Museum and Training Center, 2014

是表达建筑形式的一个条件,内化到整座建筑中去呼应建筑的内涵,可如今,这种呼应关系消失了。"在建筑形式上,由于这种形式表达"终于消失了","这样"设计重点才会放在材料被形式化之前的层次,也就是说要突出的不是材料在建筑外形上的表达,而是在这些材料被固定成某种形态之前尚未分化的状态。"[1]然而,尽管材料的物理特性营造的感知效果有些相似,但是在调动感官的维度上还是有很大不同的。如果说大地屋宇系列是调动了人的全部感官去感受大地原生态的质朴,那么利用各种技术表现金属材料物理特性的那些建筑,则仅限于对人们视觉上的调动。最后,有一点需要注意的是,在材料数据化实验盛行的现代,材料的物理特性必然局限于建筑表皮效果的表达,这是因为当材料的物理性质在应用上不是出于对建筑构造的考虑之时,这种物理性质也就只能用在装饰方面了。

承孝相让自己在物欲之外流浪,赵秉洙把自己"困"在土地之上,而郭熙秀与他们都不一样,他像一名冲浪者,跃入现实的浪潮之中。承孝相的建筑挖掘思想的深度,赵秉洙的建筑溯源原始的感觉,而郭熙秀则是追求一种现实主义的建筑外表。在他的建筑世界里,没有苦心孤诣地挖掘建筑表皮的哲学内涵或坚守一种土地信仰,对他来说,重要的是做一场建筑外观技术的"表演秀",这场秀的规则就是要吸引眼球,对于无法吸引目光的图像则要毫不留情地丢弃:"郭熙秀的建筑以夸张的造型和清晰的城市叙事风格吸引路人驻足观看。"前两位建筑师为了追求哲学的造诣或原生态的展现尽量避免炫技,正相反,郭熙秀的设计则是要尽技巧之能事。因而我们可以看到禁欲与纵欲两种不同的美学形成了鲜明的对比。

我们可以很容易在郭熙秀的建筑造型中发现夸张的元素。他设计上的首要任务就是营造一个吸引视线的空间布景。郭熙秀所采用的技术可以概括为"双景视差效应"。所谓的视差就是"从不同的角度观察物体时,在不同的位置和方向上所产生的视觉上的差异"。这种视差效应一方面会表现出时间的流动感,另一方面会显示与各个景物相关联的变化。郭熙秀的建筑设计之所以能成功地把这种视差效应赋予建筑,是因为他满足了一个前提,那就是他没把建筑当成一个整体,而是把它分割成多个单元。虽然大多时候

expressing form or space. The reason is historical. In other words, "the correspondence between the semantic system and the material foundation that has been internalized within and filled the form, as a condition of the form language, disappeared." In architecture form, for the form language, "finally disappeared," and then, in architecture form, "attention has been paid to the subcondition, that is, at the layers of the material in the undifferentiated state before being fixed in form, rather than at the level of form."[1] However, although the physical properties of the material are similar in terms of developing a phenomenal effect, the dimensions of the imported senses are very different. If the primordial sensation associated with the earth has an omni-sensory dimension, the sense of physical property combined with metal technology is almost limited to sight. Lastly, it is important to note that the physical properties of materials are inevitably limited to the surface effects when digital material experiments are active. This is because physical properties that have lost their association with the tectonics may be noting but an ornament.

Heesoo Kwak, who is set apart from the one who exiled himself from the world's desires (Seung, H-Sang) or the other who is trapped in the earth (Byoungsoo Cho), jumps into reality like a surfer riding the waves. While Seung's architecture is based on the depth of mental thought and Cho's is based on the primordial sensation, Kwak affirms the surface of reality. In the world of surface which has no depth to dig behind or support underneath, important is the play of the operational level. The rule of play here is to catch the eye because images that can't catch the eye are discarded ruthlessly: "Heesoo Kwak's architecture provides an exaggerated gesture and a legible narrative about the city so as not to pass by." While the first two architects carry out a strategy of reducing the architectural technique in pursuit of depth and primitiveness, on the contrary, Kwak adopts tactics that amplify it further. There is a sharp contrast between the aesthetics of abstinence and the aesthetics of excess.

We can find easily the exaggerated gestures in Kwak's architecture. The first task in his design is to generate a scene that captures the sight. The techniques employed by Kwak are condensed into the two-stage parallax effect. Parallax is defined as "the difference in position or direction when an object is viewed from different points". On the one hand, this shows movements that occur over time, and on the other hand points to relational changes caused by multiple

U形度假屋，2016年
U Retreat, 2016

机张Waveon咖啡馆，2016年
Gijang Waveon, 2016

他只是用了混凝土这一种材料，但他从来没有把建筑的体量看成一个整体，而是把它细分为几个部分，在平面或空间上形成一种转折（如42路住宅、莫肯民宿、U形度假屋），同时用悬挑结构把部分空间抬到空中（如F.S.One露天酒店、安阳洞"最为亲和的教堂"），从而形成视觉焦点。

  第一个视差效应在设计上可以被瞬间感知到。一些设计元素在两个视点之间反复重叠或在一个视点上并列堆叠，这样人们只要稍作移动就能观察到各种元素的交相变换。因为元素之间变化的瞬时性很强，所带来的视差效应也十分强烈，从而使这种并列和堆叠的布置不会显得松散。营造元素间极具张力的空间关系是郭式建筑的一个核心特点。与第一个瞬时性视差效应不同，第二个视差效应是在时间的流逝中被感知的，着重对比建筑不同部分随时间变化而展现的独特风貌。如果人们最终可以看出每个部分和整体规划之间的关系，这个视差效应就成功了。对比越大，效果越强。光滑表面与建筑体型的相互映衬，空间开放性和封闭性的相互变换，密实性与空透性的相互交错，地面或平台与悬挑部分的相互对比，正面和侧面、竖直方向和水平方向的各种变化，以及许多斜屋板的交织叠搭，这种全方位的变化就是整座建筑在设计规划上的关键。

  经历了15年的实验和雕琢，郭熙秀把他之前表达的各种建筑语言一并糅进了最近的这个作品"坡州庭院"中（第58页）。你可以找到许多之前的变化性元素，如位于楼梯下面的斜屋板（参考了42路住宅和F.S.One露天酒店）、混凝土亭廊和室外观景平台（参考了机张Waveon咖啡馆）、同时用作楼梯和看台的组合式空间（参考了机张Waveon咖啡馆、U形度假屋、F.S.One露天酒店、汉江Guardiana A），还有利用斜屋顶造景连通室内外的设计（参考了42路住宅、瑞文戴尔旅馆）。和坡州庭院设计最吻合的一个典型案例就是汉江Guardiana A（2009年）了，它把室内和室外空间结合起来，相互贯通，同时用一个单层建筑体量把院子围起来，再依靠楼层的交错分出层次。

objects. The prerequisite for Kwak's architecture, which creates a parallax effect, is not to treat the architecture as a monolith but to divide it into multiple elements. He mostly uses only a single material, concrete, but never treat the volume as a single volume. In his architecture, the volume is subdivided into segments, making a turn in two or three dimensions (42nd Route House, Moken Pension, U Retreat) and lifted up by a cantilever (F.S.One, The Closest Church) to become a visual focal point.

The first parallax is the effect of being perceived instantaneously, for a very short time. When architectural elements are overlapped back and forth or juxtaposed within an overlapping frame at one point in time, a shift change in the relationship between elements is perceived only by a slight movement. As the momentary relationship changes rapidly, the parallax effect is dramatic, so the relationship between overlapping or juxtaposed elements is not loose. Generating a tensed relationship between elements is the core of Kwak's architecture. The second parallax is perceived by contrasting the characteristics of one part with another by lapse of time, and is effective when it is finally possible to perceive parts in relation to the whole organization. The greater the contrast, the stronger the effect. Overall variations in smooth surface vs. volume, open vs. closed, solid vs. void, ground or flatform vs. floating volumes, front vs. side, vertical vs. horizontal, and interweaving slanting floors are the clues to organize the whole.

"Paju Gardenus (p.58)" is a collection of Kwak's architectural languages that have been experimented with and refined for 15 years. There are many variations – a sloped line below the stair volume (42nd Route House, F.S.One), concrete pavilion and outdoor viewing platform (Gijang Waveon), space created through the combination of staircase and bleachers (Gijang Waveon, U Retreat, F.S.One, Hangang Guardiana A), sloped rooftop landscaping and circulation (42nd Route House, Rivendell). The closest typological precedent is the Hangang Guardians A (2009), which combines internal and external spaces and circulation at once, while a single layer volume enclosing the courtyard is stratified on alternate floors.

The most striking element of Paju Gardenus is the courtyard. First of all, the boundary surrounding the courtyard is loosely open except for the northern side of the building. On the west side are only sloped staircases, so the first and second floors are half empty. In the south are spaces on the second floor only, so the first and third floors are outside,

汉江Guardiana A, 2009年
Hangang Guardians A, 2009

坡州庭院, 2019年
Paju Gardenus, 2019

坡州庭院最抢眼的就是院子的设计。最明显的就是，除了庭院北侧，院子的边界十分松散，向外敞开。西侧只有斜坡式楼梯，也就是说一层和三层都有一半的空间是缺失的。南侧空间上只保留二层的存在，所以一层和三层都是露天的，而东侧的空间则由二层和三层各占一半。庭院内部可供游客上楼或观览，外部可供他们在局部区域散步，或者在二层和三层上下通行。游客们在庭院内部和外部频繁地穿梭过程中，很好地感受到了整个庭院呈现的是一种强烈的立体感，而不是封闭的平面感。坐在院子里的时候，五个环绕式的楼梯映入了眼帘，楼梯下方的三个斜面朝向不同的方向，进一步增强了庭院的立体感。游客们在院子里进出时，在二层和室外的三层信步时，在楼梯间上下穿梭时，都与空间的立体性产生了各种联系。内墙和玻璃的设置更是加强了空间的这种动感。在每个地方，人们都可以根据自己和空间的这种联系来确定自己的位置和移动方向。视线始终朝向室外而不是室内。朝向东、西、南侧的楼层是露天的，同时也形成了一种空间的边界，因此视线可以向远处延伸到外围的绿色景观上。在三层的混凝土亭廊中，由平床（一种低矮的木凳）构成的全景图把整个空间的视觉感受推向了高潮。

郭熙秀似乎没有打算把结构的走向清楚地显现出来。门厅处的两根柱子与外墙呈不同的角度，到了二层的阳台上还能看到其中一根，而另一根则被内嵌到墙壁中，在三层又重新露了出来。在用一种复杂的方式做完结构设计之后，没人愿意再忠于结构本身的表达。他没有从构造角度把结构理解为建筑表现的媒介，也没有视为支配设计逻辑的因素。以他的理解，结构似乎只是一种解决问题的从属手段。

如果我们将郭熙秀的建筑分为三个阶段（第一阶段以清潭洞Tethys、42路住宅为起点；第二阶段从莫肯民宿、F.S.One露天酒店、U形度假屋开始发展；第三阶段以机张Waveon咖啡馆为标志），那么坡州庭院应该符合第三阶段。对于建筑师来说，机遇和危机是同时存在的，他们满怀期待地迎接源源不断的挑战，从而开拓进取、提升自身优势，而不是坐享自己之前的成果，止步不前。

and in the east are spaces on half of the second floor and third floor. Inside, visitors can go up and walk around the building, and outside, they can make a partial tour or climb up and down the second and third floors. The courtyard is perceived three-dimensionally, not flat, due to the frequent movement between inside and outside. When sitting in the courtyard, the five surrounding stairs are visible at a glance, and the three inclined lower plains of the staircases face in different directions, further enhancing the three-dimensionality of the courtyard. Visitors move inside and out, promenading the second and third floors outdoors, and ascending and descending the floors, making various contact with this three-dimensional space. The placement of the interior walls and the glass accelerates the dynamics of the space. Everywhere, one's position and direction are grasped according to the relationship with this three-dimensional space. Eyes are facing outward rather than inward. The outdoor floors to the east, west and south are open, while they create spatial boundaries, so the line of sight extends far into the surrounding green landscape. It reaches the climax with the panorama from the Pyeongsang – a low wooden bench – in the concrete pavilion on the third floor.
Heesoo Kwak seems not to express the flow of the structure clearly. One of the two columns standing at a different angle against the outer wall in the entrance hall remains on the second floor balcony because the other is absorbed into the wall. However, on the third floor, It reappears as a column. While solving the structure in a difficult way, there is no willingness to faithfully express the structure itself. Unlike the tectonic position, which sees structure as a mediator of expression or factor that governs the logic of organization, he seems to understand it only as a subordinate means of solving problems.
If we divide Kwak's architecture into three phases (Phase 1: sprang up at Tethys, 42nd Route House; Phase 2: pushed away at Moken Pension, F.S.One, U Retreat; and Phase 3: bore fruit at Gijang Waveon), Gardenus seems to synthesize Phase 3. The architect's opportunities and crisis are embedded at the same time. We look forward to ceaseless provocation to open a new aspect by sharpening the edge, rather than a relaxing enjoyment of the rich fruits already gathered in.

1. Mannyoung Chung, 'Again, from the Virtual to the Substantial,' *Plus*, August 1995 / 2. Mannyoung Chung, 'Superficial Narrative', *Architect*, May/June 2008 (No.232), p.11

河阳无鹤路教堂
Hayang Muhakro Church

Seung, H-Sang

河阳无鹤路教堂
Hayang Muhakro Church

Seung, H-Sang

教会的拉丁文单词是"Ecclesia",意思是那些被召唤的人。在新约《以弗所书》第1章22-23节中,圣保罗说:"上帝将万物都放在他的脚下,任命他为教会一切事务的负责人,教会是他的身体,充满着无所不在的他的一切。"

教堂被描述为基督的身体,信徒被描述为基督的四肢。因此,把教堂仅仅看作一座建筑物是错误的。教堂被理解为一个人的身体,而不是一种结构物,也就是说教堂不存在所谓的典型性建筑特征,这正是因为教堂是一群人的集合地,他们遍布世界各地,受到上帝的感召,聚集到一起。

收到召唤后就要让自己离开尘世。耶稣就是这样被上帝召唤,让自己脱离凡世,达到永生的。殉道也许是一种最为强烈的信仰行为的表现,但那些不能殉道的人必须过一种纯洁、诚实和顺服的生活,摒弃世间的贪婪和欲望,这样他们起码可以效仿基督(遵主圣范)。

无所不在的上帝不仅存在于教堂的建筑中:教堂不仅是上帝的家,也是那些被召唤者的家,那些从世俗欲望当中逐离之人的家。因此,无鹤路教堂的形式被极简化,只保留它的本质。

保罗·瓦莱里说过,最大的神秘莫过于清澈明晰。建筑师解释道:"创造一个清晰简单的空间,帮助人们寻找自由和领悟真谛。这便是这个项目的指导原则,在设计过程中这个原则始终印在我的脑海里。"

The Latin word for the church is "Ecclesia", meaning those who are called. In the New Testament, Ephesians 1:22-23, St Paul says that "God placed all things under his feet and appointed him to be head over everything for the church, which is his body, the fullness of him who fills everything in every way."

The church is described as the body of Christ, and the believers as the limbs of that body.
So it is wrong to consider the church as merely a building. Understood as a body of people, rather than a structure, no fixed prototype or typology of a church can be said to exist, precisely because the church is a community of people, across the world, who are called and gathered together by God.

Being called is to exile oneself from the world. Jesus Christ was thus called, becoming immortal by exiling himself from the world. Martyrdom may be the strongest act of faith, but those who cannot be martyred must live a life of purity, honesty, and obedience, abandoning the greed and desire of the world so that they can at least imitate Christ (Imitatio Christi).

An omnipresent God does not exist only within the confines of a church building: the church is not only the house of God, but the house of those who have been called, those who have been exiled from the desires of the world. Thus, Hayang Muhakro Church was designed in the simplest of forms, retaining only its essence.

Paul Valéry said that there is nothing more mysterious than clarity. "A clear and simple space, the house of those who search for freedom and truth – this was the guiding principle that ran through the landscape of my mind while designing this church", explained the architect.

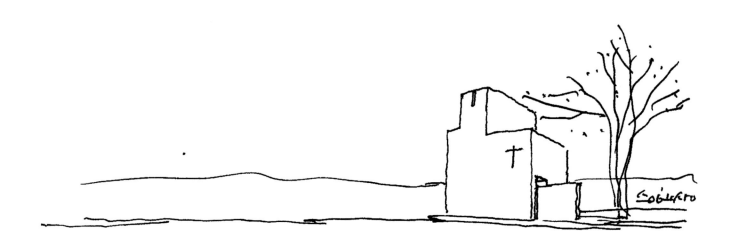

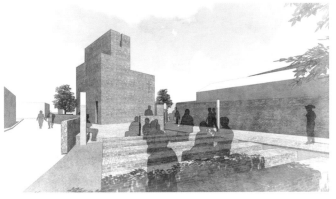

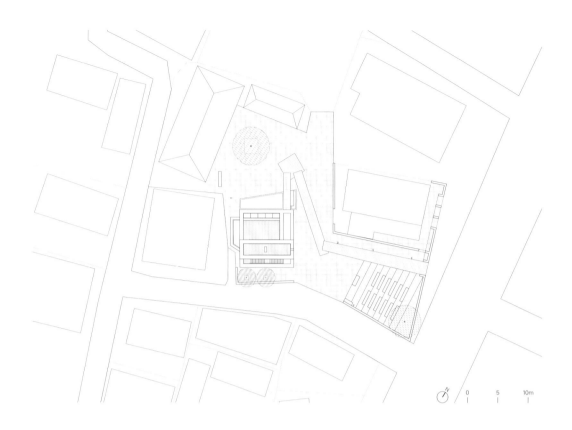

项目名称：Hayang Muhakro Church / 地点：9-4, Muhak-ro, Hayang-eup, Gyeongsan-si, Gyeongsangbuk-do, Korea
建筑师：Seung H-Sang, Lee Dong-Soo, Kim Sung-Hee / 施工方：design m, Lee Byoung-Hae / 用地面积：660m² / 建筑面积：70m²
结构：reinforced concrete / 外墙饰面：face brick / 施工时间：2018. 5—2019. 9 / 摄影师：©JongOh Kim (courtesy of the architect)

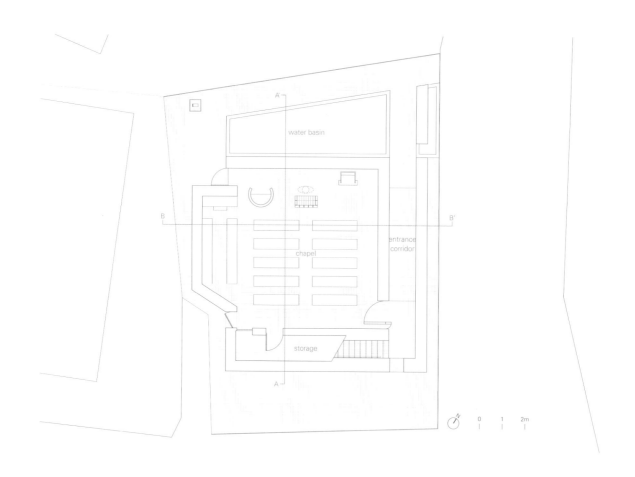

东立面 east elevation

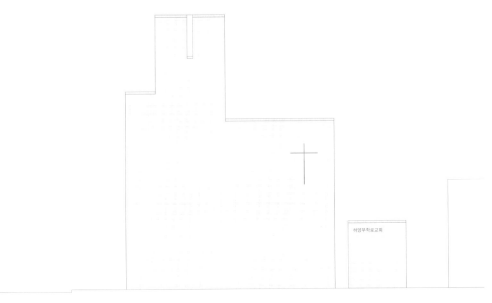

A-A' 剖面图 section A-A'

B-B' 剖面图 section B-B'

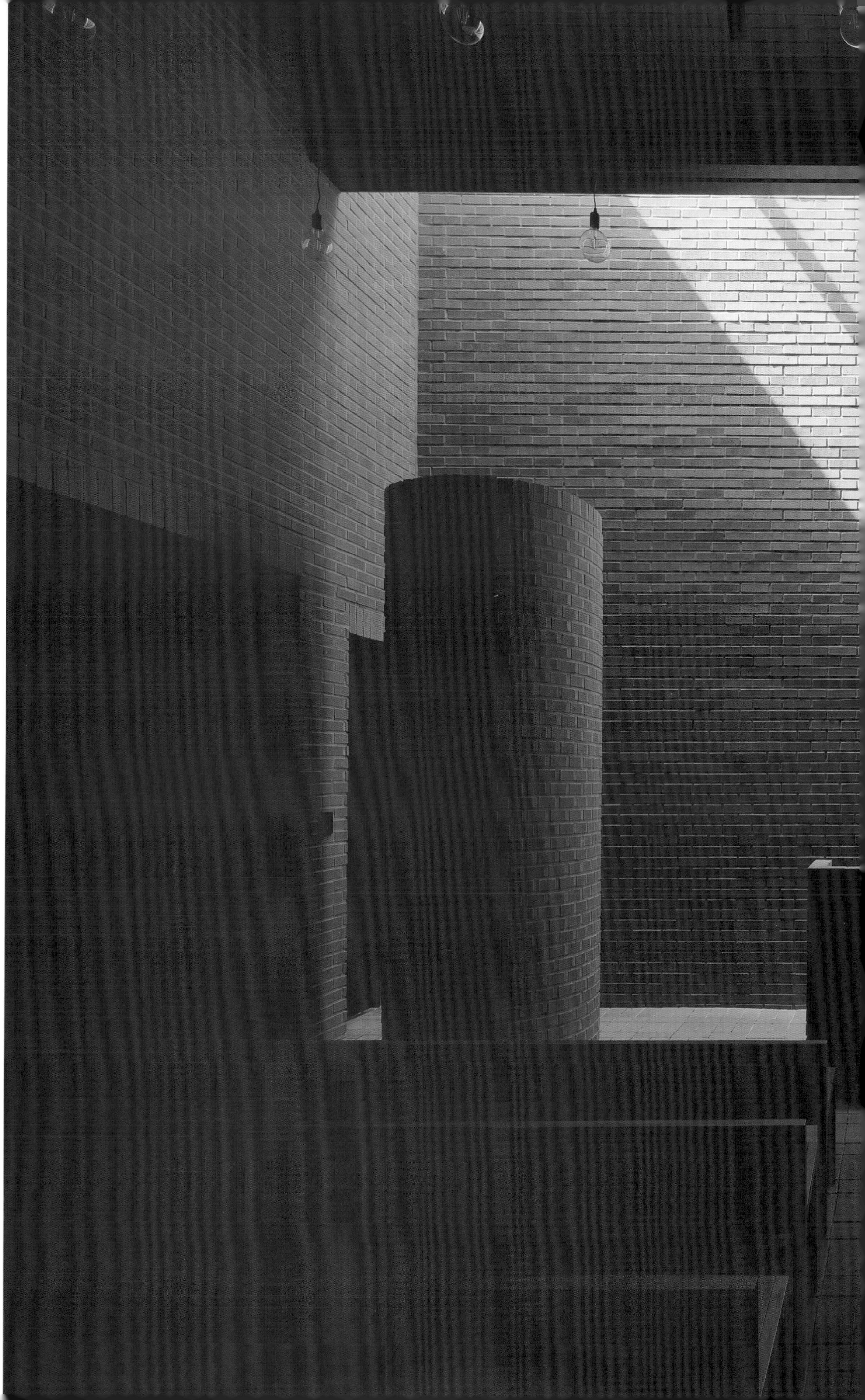

游走在张弛之间：对节制的痴迷与经历救赎后的自由
Between the obsession with temperance and the freedom of redemption

Hyon-Sob Kim

我想看看当我不再执拗于表现这种节制的时候，我会设计出怎样的建筑。¹
贫无之美……虽然缩小了我的创作范围，但实际上在这个范围内，我也有充分的自由。²

"贫无之美"总是与承孝相的作品联系在一起，而河阳无鹤路教堂似乎就是这种美的化身。在贫无之美四个字背后可能有一个"美丽"的故事，说的是韩国最好的一位建筑师，为"贫穷"的他乡异客们设计了一座教堂，"无"设计费。

在一家日报的一篇多少有点夸张的报道之后，这座教堂成了全城热议的话题。教会的成员说，"贫无之美的概念始于1992年守拙堂的设计，在无鹤路教堂中臻于完美。"可以看出他并没有说贫无之美的渊源与教堂完全无关，因为这个教堂就是在至深的虔诚中以及贫困的情况下诞生的。

但是，如果我们必须把"贫无之美"纳入我们的解释，一个更合理的理解就是在最大限度减少冗余设计的小型教堂中挖掘"贫无之美"，并对这个核心的设计理念本身进行探究。如果是的话，那么"贫无之美"到底是什么？

如果我们接受这一设计理念（尽管对此有部分人反驳），那么它所蕴含的贫无的浪漫并不是因为先前这个地方恰好也是一种贫困的状态，而是在节制和博爱之间，将灵魂填满的张力所带来的精神愉悦。

它是一种从贫穷而纯粹的生活中收获的灵魂之美，即能够安于贫穷的生活，以诚实的表现为乐，此乃正直的学者风范。也许距离理想的实现还要走一段路，但最好把它看作是到达目的地之前的一段苦旅，关注在清心寡欲的状

When I am freed from this obsession with temperance, I want to see my architecture at that time.¹
The Beauty of Poverty ... Although a fold of my own creation, I was in fact very much free within it.²

The "Beauty of Poverty", which is always attached to the work of Seung, H-Sang, seems to be personified by Hayang Muhakro Church. This may be due to the 'beautiful' story that one of Korea's best architects has designed a church "for free" for a "poor" outlander church.

After a somewhat exaggerated report in a daily newspaper, this church is the talk of the town. A member of the church says, "The 'Beauty of Poverty', which began with Sujoldang in 1992, has been completed in the Hayang Muhakro Church" - a comment not entirely without reason on the part of the church, since this new church building was born out of deepest devotion as well as needy circumstance.

But if we have to bring the "Beauty of Poverty" into our interpretation, a more appropriate understanding is to find the "Beauty of Poverty" in the correspondence of the conditions of a small church, that minimizes redundancy, and the core thesis itself. If so, what is the "Beauty of Poverty"?

If we accept this thesis despite some counterarguments against it, it does not simply imply a sterile romance which happens to have been preceded by a state of poverty, but rather it is a lyrical feeling of '"spiritually-filled tension", between temperance and fraternity.

It is the beauty of the soul acquired through leading a poor and pure life, being content in poverty and taking pleasure through acting in an honest way – the behaviors of a scholar of integrity. Perhaps the completion of the ideal may be still some way off, but it is better to see it as a journey towards the goal. The key is in appreciating how much composure, or freedom, can be enjoyed in the tension of temperance.

守拙堂，1992年
Sujoldang, 1992

态下，人们可以享受到多少的平静和自由。

　　这座教堂位于庆尚北道庆山市河阳县。室内空间约7.5m×7.5m，最多能容纳50人。教堂外前面的诵经台和仪式台都是用砖砌的，旁边放了布道者的椅子。左右两排各摆5行长椅作为观众席，唱诗团座椅有两行，在观众席的左侧，与观众席座椅方向垂直。

　　教堂内部空间的安排非常朴实无华，阳光从布道台上方的天窗射入，照亮了整个室内空间。另一方面，这座教堂的外型是由箱子的空间几何特征决定的。通过改变各个墙体的高度，设计师既克服了这种几何造型可能带来的单调感，又保留了它的几何意义所产生的情绪张力。尤其是在偏黄棕色砖砌外墙上形成的那一道道线性的纹理，堪称一绝。由于得到了大邱市的一家砖材厂的捐赠，砖块自然成为决定教堂整体印象的元素。和建筑物外部一样，里面也都是砖砌的饰面。整座教堂各个部分都使用了单一的材料——砖展现了节制之美，减少了钢筋混凝土结构的使用，并对部分使用的地方全部做了遮挡。

　　建筑师承孝相想要表达的是，在这种节制的空间里可以看到材料细微的变化。天花板的五个拱跨改变了整个室内的空间感。几个世纪以来，教堂内部都是由圆形或拱形的穹顶覆盖，象征着天堂和神明。承孝相的尝试可能并没有十分独特，然而有趣的是，这种传统做法通过砖材被应用到了当代韩国教堂的设计中，让人联想起瑞典建筑师西格德·莱韦伦茨在20世纪60年代设计的圣马克教堂和圣彼得教堂。河阳无鹤路教堂虽然内部空间不大，但是砖墙、

Located in Hayang-eup, Kyungsan, Hayang Muhakro church, which measures 7.5 meters on both sides, contains only 50 seats. At the front, the lectern and service-structure are made of brick, and a preacher's chair was placed beside them. The five, long, stall-chairs were laid out in two rows, and on the left-hand side, chairs for the choir are rotated 90 degrees and arranged in two rows.

The interior space is designed in a very modest way. Light from the skylight on the ceiling above where the sermon takes place, illuminates the church. On the other hand, the exterior of the church is dominated by the abstraction of the geometrical box. The possible monotony of the abstraction has been overcome by changing the height of the walls, while the tension from the geometrical abstraction is retained. It is noteworthy that this external lineament is filled with the material of yellowish-brown bricks. Donated from a brick factory in Daegu, this brick decides the impression of the church as a whole. The inside was also finished entirely with bricks. So this church demonstrates the beauty of temperance by using only a single material – brick – for everything to save the reinforced concrete structure, partly used and wholly concealed.

The architect Seung, H-Sang wanted to express a nuanced change in the moderation of the material. It is the five-vaulted ceiling that gives the change to the sense of the interior volume. For centuries, religious spaces have been covered by domed or vaulted ceilings, symbolizing heaven and divinity; Seung's attempt might not be that unique. However, it would be interesting to notice that such a tradition was applied in a contemporary Korean church, through the material of brick. It is reminiscent of Swedish architect Sigurd Lewerentz's St Mark's Church and St Peter's Church of the 1960s. Hayang Muhakro Church has a small interior space, but the brick walls and vaulted ceilings, skylights and textured furniture increase the density of the interior.

Just as there is a nuanced change in both the external body and the internal volume of the church

圣马克教堂，西格德·莱韦伦茨，瑞典斯德哥尔摩，1960年
St. Mark Church by Sigurd Lewerentz, Stockholm, Sweden, 1960

    拱形天花板、天窗和粗糙不平的桌椅摆设增加了室内空间的密度。

    正如教堂内外空间在体积和构图上都有微妙的变化，包围整个建筑内部的空间结构也不是一成不变的。最有特点的是，通往入口处并介于两道狭长的墙壁之间的小路，为进入教堂做好了思想准备，并赋予空间在走向上的变化。这条小路通向教堂后方狭窄陡峭的楼梯，小路和楼梯都为教堂内部的单层空间增加了一层空间。在这里，通往屋顶祷告室的楼梯就像一条苦伤道（通向上帝的忏悔之路）。

    虽然室内空间大约只有50m²，但是承孝相充分利用了室外空间，为教堂争取了一大片区域。首先，长方形的祈祷室占据了屋顶的三分之一，它没有屋盖，向天空敞开。尽管如此，整个祈祷室仍然像一个单独的房间，因为整个空间被高达4m的墙体包围起来，形成一种房间的感觉。1980年，他在老师金寿根门下学习期间，参与了庆东教堂开放式屋顶的设计，这个想法可能来自于那时候的经验。

    此外，他在教堂前院布置了一个室外教堂，包括一个讲台和一系列的长椅，扩大了教堂活动的空间。这让人联想到勒·柯布西耶廊香教堂的一个室外小教堂，它也是以其中一个立面为背景，但不同的是，在这里有精心布置的长凳。沿着场地边界建造一些低矮的栅栏，这个室外空间就被这样确定了。地面上相邻的图案将人带往前头提到的小路或是教堂的门口。与教堂密集的结构布置相比，地面上的图案就会显得松散很多，使人们在感受到教堂节制与紧张的空间感之前，先得到了一定的放松的感觉。

amid the scale and composition, so the spatial structure of the whole building enveloping the interior is not monotonous, either. Above all, the path to the entrance, which follows the long wall, readies the mind to enter the church, and gives variety to the flow of space. This path leads to a narrow and steep stairway that is set in the rear of the church, both the path and the stairway add one more spatial layer to the one layered space inside. Here, the stairway, ascending to a prayer space on the rooftop, is like the Via Dolorosa – a penitential route towards God.

Although the indoor space is limited to about 50m², Seung H-Sang took full advantage of the outdoor space and secured a large area for the church. First, the long rectangular-shaped prayer space occupying about one-third of the rooftop, without the roof, feels like a single room though it is open to the sky. It is a 4m-high wall surrounding this space that, creates the feeling of a room. This idea may have come from his experience under Kim Swoo-geun in designing an open chapel on the rooftop of Kyungdong Church in 1980.

In addition, Seung, H-Sang placed an outdoor chapel – consisting of a lectern and a series of benches in the front yard of the church, expanding the space to be used for church activities. This is reminiscent of an outdoor chapel at the Notre Dame du Haut in Ronchamp by Le Corbusier, which uses one of its facades as a background, but differs in the presence of the well-arranged benches. A low fence built along the site boundary defines this outdoor space. The adjacent ground pattern leads to the above-mentioned path or doorway of the church. Compared to the dense structure of the building, this ground pattern is considerably loose. It creates a feeling of relaxation before reaching the space of temperance and tension.

"Beauty of Poverty" reveals the craving for inner fulfillment through temperance and tension. In other

河阳无鹤路教堂的室外小教堂
outdoor chapel of Hayang Muhakro Church

"贫无之美"反映了他渴望通过设计上的节制与张力达到内在的一种完善的想法。换句话说，"贫无之美"的概念自20世纪90年代以来，一直是承孝相的建筑语言，因此这座教堂的设计也是如此。然而，他在最近接受作者的采访时表示，"贫无之美"的内涵比以前更加灵活了。除此之外，更大的不同自然是承孝相如今的国际地位了。

正如我们仔细观察到的，河阳无鹤路教堂凭借着几何造型上的抽象含义和清晰明确的空间构图表现了节制的美感，与此同时还带有一些变化的元素和舒缓压抑的元素，也就是形成了一张一弛间的自由。但是，很明显，这里的自由有别于"艺术家们创作上的自由或人们表达欲望的自由"。就像基督教悖论的"自由"一样，必须在十字架上经受了苦难才可享有救赎的恩典。对某些人来说，这种节制做得有些过分了——似乎作为基督徒的承孝相本人在年少时就一直没能摆脱对这种节制的强迫观念。但是他承认，当他回顾"贫无之美"时，发现自己被解放了，因为他曾经被限制在那里。如果我们接受他在本文开头的观点，那么承孝相的设计生涯已从对节制的痴迷转变为经历救赎后的自由。

然而，这种自由对承孝相来说是位于精神深处的、无形的、非常个人化的东西——很难看出它是如何与建筑的空间和形式直接联系起来的。任何人在任何建筑、任何条件下面对真理，都将拥有无限的自由。另外，他的自由是否能够创造出他在勒·柯布西耶的作品拉·图雷特修道院中所体验到的张力和激情？让我们为在这个小教堂里寻求真理的人们祈祷，愿他们找到真理，获得自由。

words, the "Beauty of Poverty" is a concept that has come to represent the architecture of Seung, since the 1990s onwards, so too with the design of this church. However, according to Seung in a recent interview with the author, the "Beauty of Poverty" is now more flexible than before - the bigger difference, of course, being Seung's position in the world.

As we have carefully observed, the Hayang Muhakro Church shows the beauty of temperance in its geometrical, abstract form and straightforward composition of space. At the same time, it embodies elements of change and of relaxation, which is to say, freedom within tension. However, it is obvious that the "freedom" Seung refers to is different from the freedom of artists or the expressive desires of individuals. When he looks back at the "Beauty of Poverty", he finds himself liberated because he was confined there. If we accept his remarks at the beginning of this article, over the course of his career, Seung's architecture has transformed from the obsession with temperance to the freedom of redemption.

Nevertheless, this freedom for Seung is deeply spiritual, intangible and extremely personal – it is difficult to see how this can be directly linked to the space and form of architecture. Anyone who confronts the truth, in any architecture, under any conditions, will have infinite freedom. In addition, will Seung's freedom be capable of creating the intensity and passion that he has experienced at Le Corbusier's La Tourette? Let's pray for the truth-seekers in this small church, that they find the truth they seek and gain freedom.

1. Seung, H-Sang, during his conversation with Jong-ho Yi on his religious architecture, "A road to freedom", *Plus*, March 1996 (No.107)
2. Seung, H-Sang. "Afterward: A Small Wish", *Sensuous Plan: The Architecture of Seung, H-Sang*, Pai, Hyungmin., (Paju: Dongnyuk, 2007), p. 552

巨济岛宾馆
Guest House Geojedo

BCHO Partners

线性的混凝土结构和敞开式的屋顶把温暖的阳光迎进室内，这栋建筑没有直面前方汹涌的大海，而是把身子谦卑地埋卧于大地中，安静地注视着大海。

建筑师试图尊重长久以来守在那片土地上的植被，修复已经受损的地貌，在这个复原的景观框架内，建造一个开口供人们居住。

建筑方案是创造一个安静的地方，在供游客休息的同时，还能把海平面和周围环境的美感凸显出来。建筑的一部分嵌入地下，以谦卑的姿态顺应复杂的沟壑走向和山谷的地貌特征，并真正与土地融为一体。建筑师希望通过这种建筑形式让人更直接地与自然沟通，对其进行探索。

虽然这个方案在很多方面类似于之前的"斜屋顶住宅""社区中心"及"大地屋宇"这些作品，但这个宾馆（包括它的景观和道路）周围的地形环境相比之前要复杂得多，并且这个宾馆在设计上需要考量的因素也比以前更为复杂。

对于业主来说，宾馆肯定是一个具有特殊意义的场所，业主就像一个母亲那样住在那里，打理整个宅子。对于来自五湖四海的旅客们来说，宾馆也同样具有特别的意义。建筑师希望采用独特而简单的环境来呈现宾馆的这种特质，以此来烘托土地本身的美感，并采用简单的几何线条和线条之间的空间更清晰地把这种美感表现出来。

With its linear, concrete forms and a roof flung open to let in the warm light, this house does not confront the rough sea ahead but instead sinks humbly into the earth, from where it looks serenely on.
The intention is to respect the local flora, to heal the changes that have been inflicted upon the land, and then – within the framework of this restored and healed landscape – to make an opening for the humans who will live there.

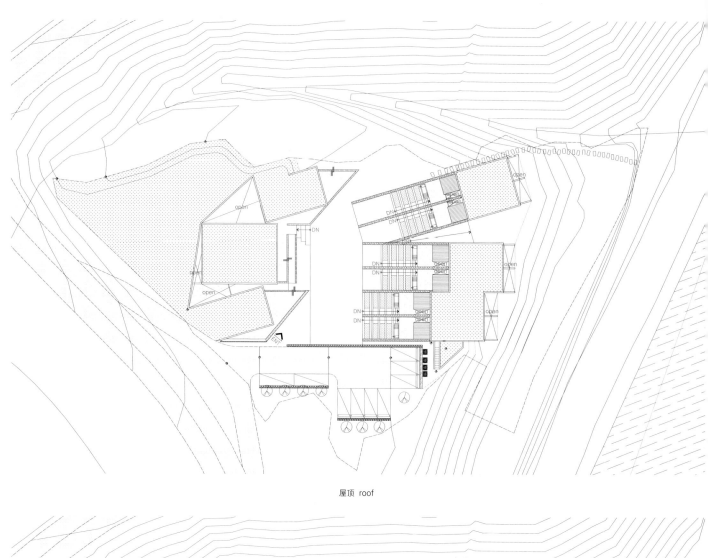

屋顶 roof

1. 宾馆A-1号楼 2. 门厅 3. 宾馆A-2号楼 4. 宾馆B-1号楼 5. 宾馆B-2号楼 6. 宾馆C-1号楼 7. 宾馆C-2号楼 8. 宾馆D-1号楼 9. 宾馆D-2号楼 10. 停车场
1. guesthouse A-1 2. lobby 3. guesthouse A-2 4. guesthouse B-1 5. guesthouse B-2 6. guesthouse C-1 7. guesthouse C-2 8. guesthouse D-1 9. guesthouse D-2 10. parking lot

一层 ground floor

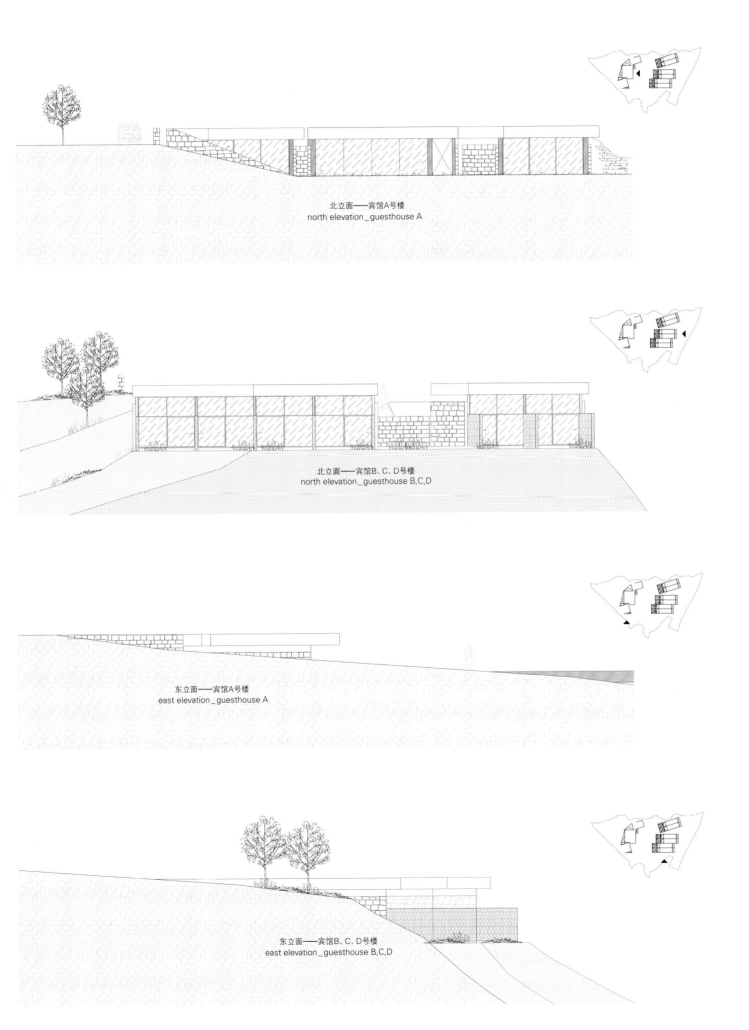

The architectural proposal is to create a peaceful place that can provide rest to visitors, whilst highlighting the beauty of the horizon and its surroundings. The architect's hope was to create a more explicit communion with, and exploration of, nature through an architecture that cuts partially into the earth, voluntarily lowering itself to follow the complex dips and valleys of the earth and essentially melting into it. Although this architectural proposal has similar aspects to the "Tilted Roof House", the "Community Center" or the "Earth House", the area that this house is set in has many more topographical complexities – such as the landscape and the roads – and furthermore, the purpose of the house is also much more complex.

The house must be a special place for the owner – a mother who will be living there and keeping house – as well as for the travelers from many different places who visit. The architect's hope is that this special quality will be provided by the unique and simple surroundings. In this way, leisure for introspection can reveal the beauty of the earth, which will become more apparent through the simple geometric lines and the spaces between them.

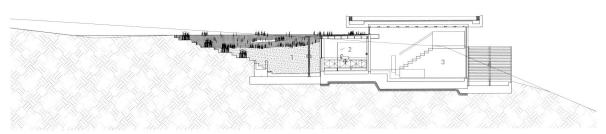

A-A' 剖面图 section A-A'

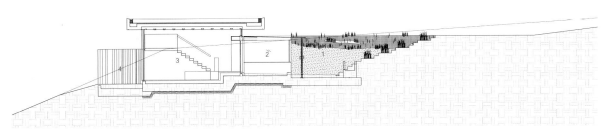

B-B' 剖面图 section B-B'

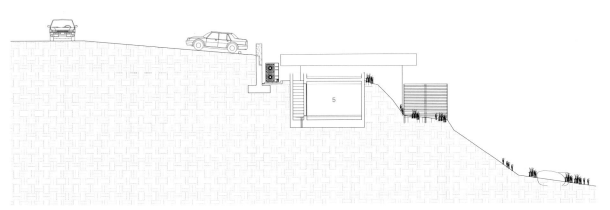

C-C' 剖面图 section C-C'

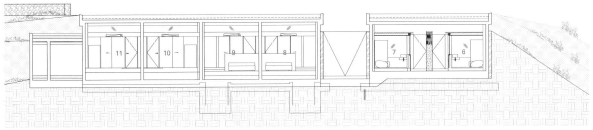

D-D' 剖面图 section D-D'

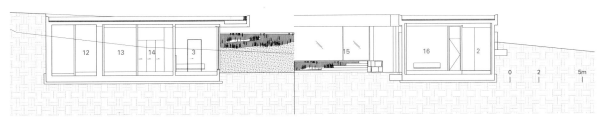

E-E' 剖面图 section E-E'

1. 后院 2. 卫生间 3. 卧室 4. 前院 5. 门卫室 6. 宾馆B-1号楼 7. 宾馆B-2号楼 8. 宾馆C-1号楼
9. 宾馆C-2号楼 10. 宾馆D-1号楼 11. 宾馆D-2号楼 12. 锅炉房 13. 厨房 14. 地热房间 15. 零售商店 16. 客房
1. backyard 2. restroom 3. bedroom 4. forecourt 5. janitor room 6. guesthouse B-1 7. guesthouse B-2 8. guesthouse C-1
9. guesthouse C-2 10. guesthouse D-1 11. guesthouse D-2 12. boiler room 13. kitchen 14. heating floor room 15. retail 16. guest room

项目名称：Guest House Geojedo / 地点：1065, Changho-ri, Sadeung-myeon, Geoje-si, Gyeongsangnam-do, Korea / 建筑师：BCHO Partners – Byoungsoo Cho, Ji-hyun Lee, Ja-yoon Yoon / 项目团队：Sook-jung Kim / 施工监理：Yoo-jin Jang / 合作方：Chaeheon construction & engineering (construction company) / 用途：guest house 用地面积：3,270m² / 建筑面积：454.57m² / 总建筑面积：454.57m² / 建筑覆盖率：13.90% / 容积率：13.90% / 建筑规模：one floor 结构：reinforced concrete / 设计时间：2016.6—2017.7 / 施工时间：2017.11—2018.9 / 摄影师：©Sergio Pirrone (courtesy of the architect) - p.38~39, p.41, p.42[lower], p.44, p.45, p.46~47, p.48, p.50[lower], p.51[top]; ©Arnold Park (courtesy of the architect) - p.42[upper], p.50[upper], p.51[bottom], p.53[lower]

巨济岛宾馆——一个与地面对话的窗口
Guest House Geojedo – a window converse with the ground

Kim Young-cheol

大地、地面和土壤；一座充满隐喻的建筑

2016年赵秉洙写道："让身体感受到土地的存在——这是可持续建筑的根本出发点，也是终极目标。[1]

长久以来，赵秉洙在建筑设计中一直特别强调"土地"的价值。

土地是什么？建筑师通常把土地解释为地面——建筑物的基础。在建筑师路德维希·密斯·凡·德·罗的设计中，所有建筑物的"座位"都被下面铺设的裙楼给抬高，创造了一个可以俯瞰周围世界的生活舞台。勒·柯布西耶的设计想要否定地面，在萨伏伊别墅中，他尽量减少建筑与地面的接触，因为他认为新的生活不可以被地面上的历史痕迹所束缚。相反，他在大西洋中发现了构建新生活的可能性：建筑必须成为一艘在海洋中航行的游轮。

许多当代建筑师把"土地"等同于"旷野"或"自然"。不论在韩国，还是在世界其他地方，"土地"已被转化成具有经济价值的资料，上面那些抽象的线条代表了各种"界限"，成为社会等级分化和权力机制的基础。

地面是一种空间概念，它不负责材料的形状或是平均特性。恰恰相反，它是用来产生材料形式和特性的源头。土地则明显具有空间和物理性质两个概念，其物理概念指的就是土壤。

土地和土壤有各种各样的指称，但很少把它们的意义放在历史文化背景下进行考虑。也许赵秉洙也有同感，所以他通过一系列作品说明人们为什么要重新认识已被遗忘的土地价值。

建筑空间的基本定义：填补旧空缺、开始新人生

工艺美学，尤其是传统工艺美学，影响了赵秉洙对建筑的看法。很多人都知道他喜欢收藏各式各样的工艺品，比如，传统油灯、缝纫工具、民间工艺品，令人惊讶的是，他的工作室里竟然还有一具人体骨架、一个孩童时期留下来的装苹果的木箱子、朋友母亲的墓碑……那些东西到现在保存多久了？现在，"土地"这个词成了他思考的对象。

Earth, ground, and soil; an architecture of metaphor

"The existence of the earth felt by the body – this is both the ultimate starting point and the destination for sustainable architecture," Byoungsoo Cho wrote, back in 2016.[1]

Byoungsoo Cho has, for a long time, placed a special value on "earth" in architecture.
Earth? Architects often interpret this idea as the ground – a base on which to construct a building. Mies van der Rohe raised the level of the seat for his buildings, placing a podium on the ground, creating a stage for life which overlooked the world around it. Le Corbusier wanted to deny the ground; in Villa Savoye, he minimized the contact of his architecture with the ground because he thought that the possibility of a new life was shackled by the traces of history in the ground. Instead, he found the possibility of new life in the Atlantic: architecture had to become a cruise ship sailing the ocean. Many contemporary architects equate "the earth" with "the wild" or with nature. In Korea, and elsewhere across the globe, "the ground" has been converted into something of economic value, abstract lines denoting "boundaries" that become the foundation for social hierarchies and the mechanisms of power.
The ground is thought a kind of spatial concept. It doesn't take charge of the form or mean property of a material. Rather, it seems to be a principle that brings them about. Earth is clearly a spatial concept as well as the physical material that refers to soil.
Earth and soil have a variety of referents, but their meanings are rarely considered in a historical and cultural context. Perhaps Byoungsoo Cho shares this criticism. Through his works he has presented reasons to restore the value of the Earth which has been so forgotten.

The fundamental definition of architectural space; a new life that occupies the emptiness

The aesthetics of crafts, especially traditional ones, informed Byoungsoo Cho's perspectives on architecture. He was known for keeping disparate items such as traditional oil lamps, sewing tools, folk crafts and, surprisingly, even a human skeleton in his workshop. A wooden apple box from his childhood, the tomb of his friend's mother … How much time has passed since then? Now the term of the earth has become the object of his thoughts.

他的建筑理论从工艺美学入手,随即转到建筑造型的逻辑上来,如Lee Oisoo画廊、Camerata音乐工作室和三盒屋等。虽然这些建筑在视觉外观上的确满足了我们情感上的期待,但结果不过是一些巨大化了的工艺品。简而言之,它们摒弃了"屋顶"这个早已被建筑界公认的要素,只是多用了一面墙取而代之:他为何如此大胆地从建筑中删除了屋顶的概念?在现代建筑的发展过程中,经过长时间的苦思,屋顶才被赋予了一定的象征意义,可以得到一些形式上的调整[2]。但是在赵秉洙这里,连屋顶这个建筑名词都被抹去了。剩下的部分只有纯粹的空间。纯空间的建筑是什么?到底是哪门子的空间呢?

他写道:"挖掘和深入大地内部的行为可能是人类与生俱来的,这与我们一直深藏于内心的本能有关。"[3]

"大地屋宇"是他与土地对峙之后的结果,这个系列成为他重新定义建筑意义的起点:

"进入大地内部的同时就是在进行一次重生,这是一次归根的旅程。你可以说,这种欲望是人类的本能,人们渴望这个最安全、最美好的地方,这是他们必须归属的地方,他们一定要回到这里。它是我们第一个家园,是我们最重要的家园,同时也是我们度过最为艰难之时的家园。"[4]

但是我们应该从字面上理解这篇文章吗? 那人类岂不是成了鼹鼠?要对此做出解释,我们必须回顾现代化的发展中对人性形象的界定:一个有创造力的孩子(尼采)、一个赤身裸体的自由自在的人(司马罗,密斯)、一个受过训练的体型健美的拳手(勒·柯布西耶)。所有这些形象都以追求自由为目标。在这个过程中,世界作为我们生活的舞台,要么留下一片空白,要么满载着物质成果,仿佛和深海里的光景一样。因此,一些人试图创造一个新的物体,而另一些人必须经历一段黑暗的世界,才能找到生存的地方,让下一代生活在光明之中。这是自希腊哲学家德谟克利特时代就存在的一种一分为二的世界观。在当代文艺理论中依然可以看到这种矛盾成分的存在。

20世纪的建筑史一直在探讨空间这一主题,而它的重要性最终得到了提升。"空间意志"这一主题以当代理论和空间形态类型学为背景,得到了各个方面的发展。

Originating in the aesthetics of crafts, Byoungsoo Cho's architectural theory moved on to the logic of architectural forms; Lee Oisoo House, Camerata Music Studio, and Three Box House, etc. Though the visual appearance of these works satisfies our emotions, the results are but largescale artifacts. Simply put, they don't share the meaning of "a roof" that has long been accepted within architecture. Instead they simply had one more wall: how dare he remove the concept of "roof" from his architecture? In the development of modern architecture, the roof was replaced with the symbolic form of itself, only after lengthy throes[2]; but even the name was wiped out in his architecture. The only thing left was pure space. The architecture of space? What kind of space is this?

"The act of excavating and going within the earth is perhaps innate in all humanity and relates to an instinct we have always carried deep within us," he writes.[3]

The "Earth House", that is the result of his confrontation with the earth, provided a starting point for his attempts to redefine the meaning of architecture:

"To enter the earth is at once a re-birth and a return to one's roots. One might say that a desire to enter the earth is instinctive for humans, yearning for the safest and best place of which humans must be a part and to which they must return. It is our first and most fundamental home and yet simultaneously the home of our greatest extremis."[4]

But should we understand this text literally? Humans are not like moles. For an explanation we must look back to the image of humanity as defined in the development of modernity; a creative child (Nietzsche), a naked free man (Schmarsow, Mies), a trained boxer of physical beauty (Le Corbusier). The goal of all these was freedom. In this process, the world as the stage of our lives was left as a tabula rasa or filled with material, like the world in the deep sea. So some tried to create a new object, and others had to go through a world of darkness to finally make a place for life, filled with light for the next generation. This is one of the dichotomized worldviews existing since the time of the Greek philosopher Democritus. This conflict composition can be found in the field of contemporary art theory.

The history of architecture in the 20th century kept tackling this subject of space, and it eventually stepped up in significance. The subject of "Raumwille", set in the modern theory and the typology of spatial formation, was developed in various ways.

然而，关于"世界观"的讨论似乎还没有开始。人们常说，一个空间被置于世界之中，空间的形式取决于它与世界的关系。那么，赵秉洙所说的这个空间到底建立在什么"基础"上呢？

我们通过"基础"回到原来的关于土地的问题上。那么，赵秉洙在这里讨论的土地到底是什么意思呢？

当他说"进入大地内部"或"重生的地方"时，他的意思是"地球是物质的本原"。这种表述，我们已经听了很久了。无论是西方的恩培多克勒的四根说，还是东方经典的五行说，地球都是世界和宇宙的法则。因此，赵秉洙的建筑依靠土地、属于土地、忠于土地。对他来说，把人为与自然分开、把理性与信仰分开，这些常见的做法是毫无意义的。一个生命把自己在地球上的世界设定为地球的存在，那它就存在于时间的地平线上。

这个世界就诞生在黑色坚实的土地上。我们还没有完全找到思考这种结构的方法，似乎别无选择只好跟着"他"走，因为每当"他"的思想经过某个地方的时候，我们就有了一处可以生活的场所。

几何秩序：赵秉洙的建筑创作逻辑

赵秉洙通过一定的几何秩序构建了他的建筑空间。在挖掘土地的过程中，他也承认了建筑史是以几何学的名义有序地记录下来的。这种几何秩序向我们展示了一个更有意义的世界。

同他之前的大地屋宇系列一样，巨济岛宾馆在设计上也被规定成了一个基本的几何形状，样子如同他的那个"苹果木箱"被嵌进了土体中。[5] 在他看来，他对苹果木箱的记忆和感觉帮助他生成了这么一个场所，物质和内涵两个维度在这里发生了本质性的碰撞。虽然原本只是一个装着苹果的箱子，但是它也能为我们和宇宙的沟通提供一个框架。因为当镶板本身"说话"的那一刻，它就与所面对的世界发生了沟通。他说这是"对于我们人类的一种定义"[6]。

我们既不是毫无意识也不是随心所欲的个体。对他来说，普遍存在的秩序呼应了我们的方方面面。他的空间总是默默地容纳着我们日常生活的点点滴滴。当我们坐在那里时，世界被"璀璨的星空"填满，它没有具体形状，只有空间的存在，成为大地、天空和自然之间的纽带。

However, it seems that the discussion of "Weltbild," a conception of the world based on subjects and forms has not yet begun. It is often said that a space is placed in the world, whose form is defined by its relation to the world. Then, what is the "foundation" on which the space stands?

We come back to the question of ground through the "foundation". So, what is the meaning of the earth when Cho now discusses it?

When he says "entering the earth" or "the place of re-birth", he means "the earth as a principle". The earth as a principle: we have heard this expression for a long time. Not only in the Western tradition of Empedocles but also in the Oriental classics, the earth was the principle of the world and the universe. It means not only the place where we are born, but also the birth of continuous life itself. Therefore, Cho is carrying out architecture by the earth, of the earth, and for the earth. To him, it is meaningless to separate artificiality from nature and divide rationality from faith as is often the case. A life to set forth one's world on the earth as the existence of the earth is the existence of the being on the horizon of time.

The world is made of the earth: a dark and solid medium. We have not yet fully figured out the means by which to think of that structure. It seems we have no choice but to follow him, because we can find a place for our life, after his thinking has passed through there.

Order of Geometry; The logic of Creation in Byoungsoo Cho's architecture

Byoungsoo Cho has realized his architectural spaces through geometric order. When excavating the earth, he also accepted that the experiences of architectural history were inscribed, in an orderly way, in the name of geometry. This presents a world to us more meaningfully.

In Guest House Geojedo, a fundamental form such as his "apple box" embedded in the ground is inherent in this house, as in his previous houses.[5] To him, the memory and sense of the apple box provides a place where the domains of materiality and meaning meet in an essential way. The box is only filled with apples, but it also provides a framework for communion between us and the universe, as the moment when the panel itself speaks, it communicates with the world which the material will face. He called this "a defining of us".[6]

We are neither spontaneous nor arbitrary individuals. The universal order for him corresponds to globality for us. His space accommodates our daily life silently. When we take a seat there, the world is filled with "a night sky with stars that shine brilliantly", and a place without form, where only space exists, becomes a connector between the earth, the sky and nature.

Ground and Sky; The Power of the Horizon

He picked up Ko Yu-sup's book[7] and opened it, revealing well-thumbed pages and releasing the deep scent of a book:

地面与天空的对话：地平线的力量

他拿起高裕燮写的一本书[7]开始翻看起来，从书页的痕迹可以看出这本书他经常翻看，从中散发出一股浓浓的书香。"我总会情不自禁地进入东方的绘画世界，那里有许多物体需要你去'识别'，你不能仅仅从物体本身的存在对它们进行辨认，更要通过画里面人的行为去识别周围的物体……因此，东方的山水画中总有人的存在。面前的画中之人不是一个物体，也不是与我对立的一个绘画者，他就是我自己的化身，我跟着他一起感受画中的风景，沿溪而下，听听鸟鸣，闻闻花香。"[8]

所以巨济岛宾馆也是他自己的场所。这片土地很长一段时间曾是农耕用地，他重新把这里变成了一个有意义的空间。他把这里的空间敞开，让它面朝大海、望向天空，成为面对世界的一扇又一扇窗口。在这里，你可以收获安静地存在于内心深处的快乐，这是一种忘我的心灵体验，不受他人意志的控制。你在这里的独处完全出于自觉，所以你不会觉得孤独，反而会获得一种精神上的愉悦，它并不是世俗意义上的快乐，也十分不同于宗教上的清规戒律。[9]

自律的美经常被称作是悲伤之美，但我更愿意把它解读为崇高之美。因为它不是出自一种悲天悯人的情感，而是一篇寻找信仰世界的心灵日记。

我不知道赵秉洙所强调的是神秘主义还是心灵直觉。不管怎样，他给我上了一课，那就是即使一个微不足道的物体也可以是积极向上的。他没有拒绝无用或徒劳的想法，他的建筑世界不仅要表现美丽和温柔，还要有敏锐、柔软和悲伤。他没有停下脚步，继续追求着建筑的几何秩序。

这座宾馆不是一个远离日常生活的地方，而是一个与大地面对面的地方。它把我们放在世界的面前，证明我们自己在这个世界上的存在。

我想起了阿西西的圣方济各曾经踏过的土地。每当天空开始下起雨时，体质虚弱的圣方济各还是欢欢乐乐、蹦蹦跳跳，他说下雨是天空和大地在举行婚礼。[10]

雨将上帝与地上的他连在一起，上帝赐给世界何等的祝福！雨滴溅起泥土跳动的芬芳也是来自上帝的恩赐。

"I cannot resist entering into the picture, in oriental painting, which requires one to "recognize" objects. Not only by existence, but also through a physical act... Therefore, in an oriental landscape painting, man always exists. He is not an object or an opposing figure as a painter who confronts me, but he is myself, able to experience the landscape, to follow the stream of water, to listen to the birds, smell the flowers."[8]

So Guest House Geojedo is also his own place. He has transformed part of the long-cultivated land into a meaningful space once again, opened the empty space toward the sea, and further toward the sky. These are windows for confrontation with the world. A place to embrace the joy that exists quietly, deep in the mind, such as self-forgetfulness, mindfulness and a relinquishing of any trace of will. A euphoria that comes from a fully conscious solitude, that is not lonely, but is far from both secular pleasure and religious precepts.[9]

Autonomous beauty is often called the beauty of sorrow, but I would rather translate it as the beauty of the sublime. Because it is not a sympathy or compassion for the life of human beings, but a documentation of the mind seeking the world of faith.

I don't know whether Byoungsoo Cho puts emphasis on mysticism or intuition. Either way, I can learn a lesson from him, that even a meaningless entity can be positive. He does not reject inane or vain thoughts; the world in which he enters presents not only beauty, and gentleness, but also sharpness, softness and sadness. As far as the order of geometry is concerned, he remains in relentless pursuit.

The house is not a place of deviation from everyday life, but it is a place of confrontation with the earth; Guest House Geojedo is architecture as a space which makes our existence confront the world, and convince us of our own existence in this world.

I am reminded of the earth which Saint Francis of Assisi used to step on. When it started to rain he used to leap for joy despite his weak constitution, calling it as the marriage of heaven and earth.[10]

What a blessing from the god to the world that is combined with himself to him who came from the earth! The raindrops that brought him the fragrance of the earth in a jump were also a divine blessing.

---

1. Byoungsoo Cho, *House in the Earth, House towards the Earth*, (Seoul: SPACE, 2016), p.18 / 2. Josef Frank, *Architektur als Symbol*, 1931
3. Byoungsoo Cho, *House in the Earth, House towards the Earth*, (Seoul: SPACE, 2016), p.12 / 4. Ibid., p.36
5. Byoungsoo Cho, 'Thoughts on the Apple Box', *+Architect 03 Cho Byoungsoo*, (Seoul: SPACE, 2009), p.210 / 6. Ibid., p.212
7. Ko Yu-sup, *Aesthetic thesis on History of Fine Arts of Joseon*, Tongmungwan, 1963
8. Byoungsoo Cho, 'Things that I Like, Things that I Have Made', *+Architect 03 Cho Byoungsoo*, (Seoul: SPACE, 2009), p.7
9. Chae Yu-Gyung, 'Yanagi Muneyoshi's Beauty of Grief found in Joseon folk arts', *Religion and Culture*, Seoul National University Center for Religious Studies, No.16, 2009, p.98
10. Nikos Kazantzakis, *Saint Francis*, trans. Kim Youngsin, (Paju: Open Books, 2016), p.279

坡州庭院
Paju Gardenus

IDMM Architects

坡州地处坡平山和雉岳山以南一带，西部以临津江下游和汉江上游为界。两条江在乌头山统一瞭望台附近交汇，流入西海。坡州嗨里文艺村位于汉江上游的低平原上，同时也是通往西海的要道。庭院正好位于嗨里商业区的中心，周围满是富有吸引力的建筑，包括咖啡馆、餐厅、儿童娱乐场所等，功能很全面。

庭院

在城市中，一个优质的场所不应该单纯地供人进行各种娱乐消遣的活动，而是更应该让人从日常生活中就能享受到乐趣。尤其是街道，不仅可以用来通行，还可以充当公共广场。嗨里商业区街景布置得十分密集，单独供游客们休息的空间相对不足，而游客们也只是简单地逛逛就找地方休息了。因此，坡州庭院在设计概念上就十分吸引游客。它在功能上一举得二：既满足了公众活动休息的需要，也给从事营利活动的商家提供了方便。在韩国，传统庭院与社区的农业实践紧密结合，不像西方的庭院只是为了欣赏自然而设计的，要进行一定的改造。韩国的庭院不仅用于社交活动，还用来举行仪式，如婚礼、葬礼和驱魔仪式——在韩国，大多数与日常生活相关的事情都会在这里发生。

桥下休憩地

庭院的东侧与韩屋画廊宽敞的花园相邻，南侧与迎月山的山脚相连。西侧面向街道，直接与庭院相连，庭院区域是街道经过扩大形成的一个衣兜状的地带。为了给院子的夜晚增一些光感，给白天添一丝清新，在院子西侧铺了一个水池，池边是熙熙攘攘的人群，他们来来往往的身影映在水池中，划过水面。

西立面由两道巨大的混凝土墙组成，到了南侧，立面开始打开，方便采光和纳凉。连通入口和院子内部的是一个狭长轻巧的结构，以及旁边与之并行的一座桥。巨大的混凝土体块仿佛飞离地面似的，十分不羁地悬挑在空中，内部完全向街道敞开，吸引行人走进建筑。这里没有沉重的阴影和城市的喧嚣。穿透建筑的光影在院子里进进出出地形成了动态的对比，仿佛在沿着院子跳着舞蹈。设计师在院子里"种"了许多"影子"，游客们顺着"影子"的分布休憩。过去的韩国人在仲夏时节喜欢聚在桥下，因为那里阴凉宜人，还可避雨。在坡州庭院中，桥下的院子重新诠释了韩国人记忆中的生活。

混凝土亭廊

庭院内部的空间各式各样，如同迷宫般的人行道、内部的小巷和口袋空间，到处都能让你看到庭院的全景。悬浮在庭院上方的混凝土体块创造了独特氛围，游客们从柜台那里拿到咖啡之后，可以去一楼的院子里，也可以沿着入口看台式的楼梯前往位于二层的休息室。

一层的院子有两个独立的外部楼梯通往楼上。北面的楼梯通往三

层，上楼时还能看到韩屋画廊的花园。南侧的楼梯则通往二层，与迎月山紧密相连。二层是由几个回廊组成的空间，游客可以透过回廊的窗户往下看，感受庭院空间的深度。

　　画廊与庭院内部的客流分开。两个高低不同的空间表现出了明暗对比和极大的纵向差异。三层有一个混凝土亭廊。当传统的亭子都在"踮着脚尖"恨不得登到峭壁上去攀看上面的风景时，坡州庭院里的亭子却在空中"筑起了巢"。

　　夕阳即将西下，混凝土墙面在日暮中泛起了红晕。游客们躺在木椅上，在亭子里尽情享受日落带来的浪漫景致。

Paju is situated to the south of Mt Papyung and Mt Gamak. It sits downstream of the Imjin River and upstream of the Han River on its western boundary. The two rivers join at the Odusan Unification Observatory and flow down towards the West Sea. The Culture and Art Village of Heyri in Paju is located in a low plain upstream of the Han River, also en route to the West Sea. Gardenus is right in the center of the business district of Heyri, which is full of attractive architecture such as cafes, restaurants, entertainment for children, and plenty of things to do.

Courtyard - Madang

A good place in the city should not only be a destination full of activities to enjoy, but should also be a place where people can enjoy daily life too. Streets, especially, are not just passages but act as public squares. Due to the dense streetscape in the business district of Heyri, the visitors lacked space to rest and simply enjoy the space and rest. Gardenus therefore used the courtyard concept as a space to attract people.

This courtyard serves a dual function – mutually convenient for those who want to use the architectural space for public use, and those who want to pursue profit-making activities. In Korea, the courtyard's function has traditionally been closely allied with the agriculture practices of the community, unlike gardens in Western countries which are designed for the enjoyment of nature, by adapting it. Korean courtyards were not solely for social activities, but also for the performance of rituals such as wedding ceremonies, funerals and gut (exorcisms) - most matters relating to ordinary experiences in Korea would take place here.

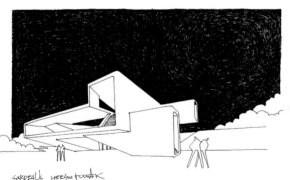

西立面 west elevation

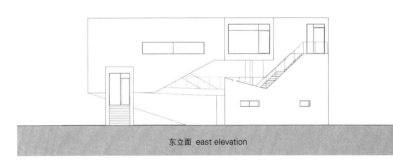

东立面 east elevation

南立面 south elevation

A retreat under a bridge

The east side of the site is adjacent to the spacious garden of the neighboring Hanok Gallery, and the south side is connected with the end of Dalmaji Hill. The west side faces the street and connects directly to the courtyard, a pocket space with an enlarged street area.

A pond is located on the west side of the courtyard to illuminate the night and bring freshness during the day. The shadows created by the people who come and go, bustling around the pond, sway across the surface of the water.

The western facade, consisting of two layers of massive concrete walls, opens up on the southern facade to pull in light and the shade of trees. Entering through the opening to the courtyard is a long, light mass and a bridge connected side by side. The concrete mass that flies freely from the courtyard opens the space toward the street and attracts people into the building. Neither heavy shadows nor the bustle of the city is to be found in Gardenus. The light and shadows that penetrate through the architecture move contrastingly in and out of the courtyard as if they are dancing along its latitude. Visitors follow the trajectory of the shadows that are implanted in the courtyard.

In midsummer, the underneath of bridges used to be a retreat space for Koreans because this was a nice, shaded place to cool off and also avoid the rain. In Gardenus, the courtyard under the bridge is a reinterpretation of this daily memory for Koreans.

Concrete pavilion

Inside Gardenus are all kinds of spaces such as a maze of walkways, interior alleys and pocket spaces as well as panoramic views. The concrete mass that is floated above the courtyard creates a distinctive atmosphere. Visitors pick up coffee at the barista counter and head out to the courtyard or up to the lounge on the second floor along the stand-like stair by the entrance.

The courtyard has two separate external stairs to the upper floors. On the north side, there is a staircase that goes up to the third floor while looking at the garden of the Hanok Gallery. On the south side is a stairway to the second floor closely attached to the Dalmaji hill. The second floor is a space made of corridors from which visitors may look down through the window to feel the depth of the courtyard space.

The gallery is separated from the flow of visitor circulation inside. The spaces of the two different heights show the contrast of darkness and light and the extreme depth differences.

A concrete pavilion was placed on the third floor. While the traditional pavilion tiptoes onto the cliffs to reach the scenery, the pavilion of Gardenus is nesting in the air.

By the time the sun declines in the west, the concrete surface turns reddish in the setting sun. Visitors can enjoy the romantic sunsets laying on the wooden bench in the concrete pavilion.

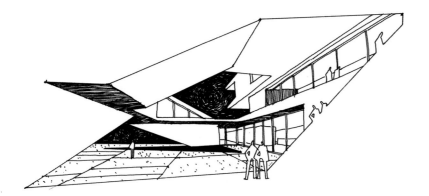
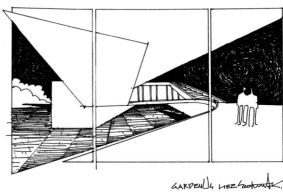

项目名称：Paju Gardenus / 地点：Beopheung-ri, Tanhyeon-myeon, Paju-si, Gyeonggi-do, Korea / 建筑师：Heesoo Kwak - IDMM Architects / 设计团队：Junsoo Kim, Jiheon Kim, Seoyoung Jang, Sunpil Hwang, Yeonju Park / 结构工程师：S.D.M PARTNERS / 机械工程师：Cheongwoo Engineering / 电子工程师：Kukdong Electric / 用途：2nd neighborhood living facility / 用地面积：1,158.5m² / 建筑面积：578.75m² / 总建筑面积：789.31m² / 建筑规模：three stories above ground / 建筑高度：11m / 容积率：49.96% / 建筑面积比例：68.13% / 结构：RC / 外饰面：exposed concrete / 设计时间：2017.6—2018.3 / 施工时间：2018.3—2018.12 / 摄影师：©Jaeyoun Kim (courtesy of the architect)

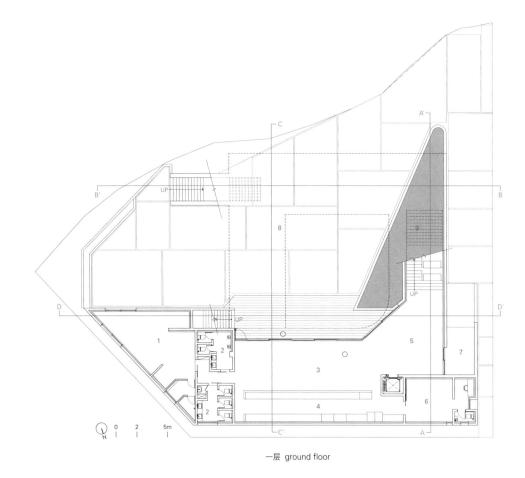

一层 ground floor

1. 面包房
2. 浴室
3. 休息室
4. 厨房
5. 大厅
6. 辅助设施
7. 主入口
8. 庭院
9. 池塘

1. bakery
2. bathroom
3. lounge
4. kitchen
5. hall
6. support facility
7. front entrance
8. Madang (courtyard)
9. pond

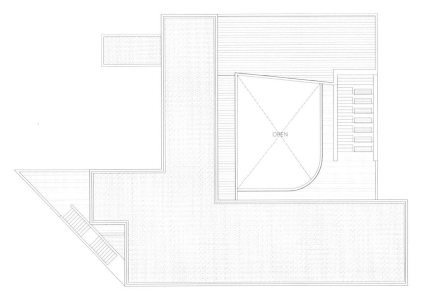

屋顶 roof

1. 长廊
2. 浴室
3. 平台

1. gallery
2. bathroom
3. deck

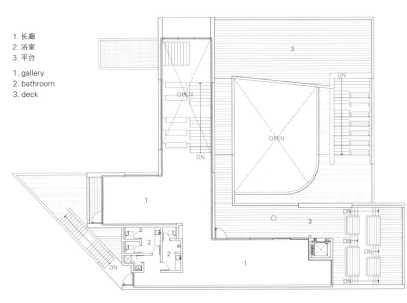

三层 second floor

1. 长廊
2. 浴室
3. 休息室
4. 平台

1. gallery
2. bathroom
3. lounge
4. deck

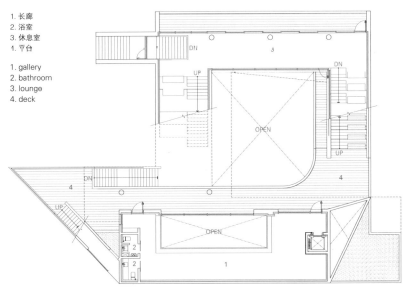

二层 first floor

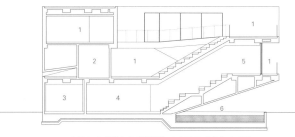

1. 平台 2. 储藏室 3. 辅助设施 4. 大厅 5. 休息室 6. 池塘
1. deck 2. storage 3. support facility 4. hall 5. lounge 6. pond
A-A' 剖面图 section A-A'

1. 平台 2. 长廊 3. 休息室 4. 池塘 5. 庭院
1. deck 2. gallery 3. lounge 4. pond 5. Madang (courtyard)
B-B' 剖面图 section B-B'

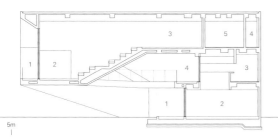

1. 平台 2. 休息室 3. 长廊 4. 走廊 5. 浴室
1. deck 2. lounge 3. gallery 4. corridor 5. bathroom
C-C' 剖面图 section C-C'

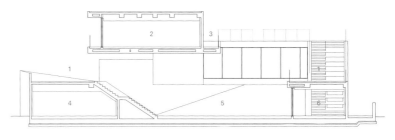

1. 平台 2. 长廊 3. 走廊 4. 面包房 5. 庭院 6. 大厅
1. deck 2. gallery 3. corridor 4. bakery 5. Madang (courtyard) 6. hall
D-D' 剖面图 section D-D'

坡州庭院：从形体到表现
Paju Gardenus: from form to behavior

Kang Howon

初夏的一个星期五下午，我来到位于坡州的嗨里村。雨季的午后，阳光充足，空气微潮。远处，一组清水混凝土结构带着锐利的线条俘获了我的目光。我一眼就能看出这是郭熙秀老师的作品。我一步步走近，一个庭院映入眼帘。简约的地板上洒着道道阴影，凉快的微风带来丝丝的清爽。

一笔挥毫的草书体
一笔书体指草书文字间笔画自始至终连绵相续，如一笔直下而成的字体。坡州庭院表现出来的就是这种一笔挥毫的美感。

整个线条从场地的东北角开始起笔，逆时针运笔，在西北角处笔锋转南，再向二层运笔，回到刚刚起笔的东北角。再继续从地面东北角开始，向三层运笔，笔锋再次转西，最后在西北角收笔，总共转了一整圈零四分之一圈。

坡州庭院连续流畅的线性造型就是沿着这个轨迹形成的。

在建筑体量的中央，有一处空地。在东南角底部加建了一个额外的体量作为二层高主体结构的支撑，东北角的三层下面也有一个支撑向东延伸，此外在北侧的一层和三层之间还有一个支撑。这些支撑构件都是线性的，宽度约5m到8m，高度为一个层高。这些混凝土支撑没有汇合，即使在竖直方向有重叠，也不做外观上的改变，就像一笔写成的书法作品一样。从平面图上看，这种结构序列十分简单：线性的建筑形体沿着场地的边界连续排列。这种空间布局也让室外空间有了立体感和连续感。每层楼的内外空间都相互连通，给人们创造了各种交流的机会。这种一笔书体的构图使空间从地面升起，让庭院的边界也模糊了起来。因此，建筑师不是强行在构图上拐弯抹角，而是借用这种一笔书的线条将庭院的结构清晰地呈现出来。

Friday afternoon, early summer in Heyri, Paju: it was sunny and humid due to the rainy season. In the distance, a group of exposed concrete volumes with sharp lines caught my eyes. At a glance, I could tell that this was the work of Heesoo Kwak. As I approached, a courtyard came into view. Shadow was cast over the simply-finished floor, accompanied by the pleasant breath of a cooling breeze.

Il-pil-seo
Il-pil-seo means a text written with one stroke of the brush. Paju Gardenus is a building that delivers the aesthetics of Il-pil-seo.

A line starts at the northeast corner of the site and whirls counterclockwise. It turns southwards at the northwest corner, and goes up one level higher, returning to the starting point at the northeast corner. Starting at the ground level, the line reaches the third floor and turns to the west one more time, finally ending at the northwest corner. In total, it completes one and a quarter turns around the site.

The continuous, linear volume of Gardenus is formed along this trajectory.

At the center of the volume, there is also an empty space. An additional volume is inserted at the southeast corner to support the main volume of two-story height, another jutting to the east after branching from the northeast corner on the third floor, and yet another between the first and third floor on the northern side.

All of them are linear; they are between five and eight meters wide with the height of one story. These linear volumes never merge, even when they overlap vertically, but they maintain their appearance, just as Il-pil-seo. The sequence looks quite simple in terms of its floor plan: linear volumes are continuously arranged along the boundary of the site. This layout makes outdoor space seem three-dimensional and continuous too, with the inside space associated with the outside on each floor, creating various opportunities for circulation. Floating from the ground, the composition of Il-pil-seo also defines the courtyard with an ambiguous boundary. It plays the role of a skeleton, clearly revealing the structure of Gardenus, rather than being simply a device that provides enforced circulation.

坡州庭院，西侧一景
Paju Gardenus, view from the west

"敞开"的院子

院子是全世界以及全人类历史上都通用的一个建筑元素。它是一个没有屋顶的空间，被墙壁和结构物包围，是室内和室外、私人空间和公共区域之间的中介空间。因为院子周围有墙把外部世界隔开，所以院子就成为用来举办各种活动的场所，但它并不属于室内空间。从定义上说，院子是一个有着明确边界的空间。

可是坡州庭院的边界似乎很模糊。三个楼层只在北侧形成了一个明显的边界。建筑东侧与韩屋画廊的花园相接，混凝土结构开始沿着东侧悬空，形成一条线，从二层升到了三层。建筑西侧的部分呈一条斜直线，从一层升到了南侧的二层。建筑从西北角到东北角的区域几乎是空的。整个结构如同在表演杂技一般，就在两个方向的悬臂空间正交的地方设置了支撑，西北角的地面几乎被留空，没有其他的支撑结构。建筑"漂浮"的线性形体婉转地削弱了边界的存在，形成了一个朝着西侧道路"敞开的"院子。

庭院向道路敞开，而所有环绕庭院的空间在各个方向上都向庭院敞开。敞开的空间照实说就是没有规定明确功能的空间，而它的存在对于密集的城市区域十分重要。虽然坡州庭院中的院子是一个私人场所，但它的公共使用价值未来仍然可期。

管道形状的变化

从外形和结构来看，一根管子只有入口和出口，在管身两侧是没有开口的。基于一笔书体的概念，我们假定这个庭院的结构是一条连续的管道，位于三层西北角的西向开口就是管道的端头，一层东南入口处的楼梯、三层东南侧的东向开口、西北侧的南

"Open" Courtyard

The courtyard has been borrowed as architectural element, all over the world, and throughout the history of mankind. It is a roofless space surrounded by walls and structures, playing an intermediary role between inside and outside, private and public. It has been used as a place for various activities, as it is separated from the world outside its walls, but it doesn't belong to the indoor realm. By definition, a courtyard is a space with a well-defined boundary.

But the boundary of the courtyard in Gardenus seems ambiguous. On the first floor, there is only one clear boundary; a volume of three-story height on the north side. A floating linear volume ascends from the second floor to the third floor along the east side which is adjacent to the garden of the Hanok gallery behind it. A linear sloping volume goes up from the first floor to the second floor toward the south on the west side, and the area from the northwest to the northeast corner is almost left empty. The acrobatic structure that supports the orthogonal volume with cantilevers in two directions leaves the ground on the northwest corner empty. Implying a weak boundary, the floating linear volume creates an "open" courtyard toward the road in the west.

While the courtyard is open toward the road, all the spaces enclosing the courtyard in three dimensions are open toward the courtyard. Eyes naturally turn to both the inside and the outside. Open space is literally that – a space with undetermined function – and it can play an especially important role in dense urban areas. Though the courtyard in Gardenus is a private space, its public use in the future remains to be seen.

Tubular Transformation

Due to their shape and structure, tubes have only two openings – at the beginning and end, not on the sides. Assuming that Gardenus is made of a continuous, tube-shaped structure, based on the concept of Il-pil-seo, the opening which faces west at the northwest corner on the third floor is the end of the tube, the stairway at the southeast entrance of the first floor, the opening facing east at the southeast on the third floor, and the opening facing south at the northeast can be regarded as the ends of branches. However, all the sides of the tube facing

"开放的"庭院
"Open" Courtyard

向开口可以看作管道各分支端。然而，管道朝着院子的所有侧面都是大范围开放的，二层南侧和与边界相邻的三层东侧也是开放的。

　　管状结构本身是可以自我支撑的，但如果其管壁被拆掉，就需要补充新的支撑了。为了做出这种侧壁开放的管状结构，郭熙秀在设计初期进行了反复的试验和纠错。南侧二层的全部侧面都被打开，这样，二层的游客们向北就能看见院子，向南则可欣赏隔壁的园景。为了给管壁上方补充支撑，只在一行上安排了3个柱子，形成了一个空腹结构。楼板下面一个柱子也没有，就像一座桥。在这里，管子的封闭结构被完全地拆掉了，可是你依然能感受到线条上一笔书体式的连续感。

从设计到行为
　　郭熙秀在之前的作品中似乎一直关注建筑本身的造型。直线、斜线、悬挑结构、一体式结构等造型元素他都尝试过，从而获得一种不拘于日常生活的动态表现。建筑虽不能飞，但是，其形体上表现出来的动态感可以让人想象，建筑似乎可以移动或飞翔。在他无数次的实验和尝试中，他能切身地体验到特定的建筑语言对人们行为产生的影响。如今在坡州庭院中，他似乎就把兴趣转到了如何通过建筑引导游客们的行动上。他在斜坡上放置了四个低矮的木质长凳，两个在里面，两个在外面。缓缓下降的坡度使躺在长凳上的人在享受舒适的同时，也能看到自己周围和前方的情况。

　　他用斜线和斜面创造出别具一格的动态形式，这些形式的设计如果不具有项目上的合理性，就很可能沦为一种纯粹博人眼球的虚荣表现。在墙壁、地板、天花板等建筑形式的基本元素中，地板对人的行为有直接的影响，它负责传递荷载、掌控人们的

the courtyard are largely open, and the south side of the second floor and the east side of third floor adjacent to the boundary are also open.
A tube can stand by itself as an independent structure but requires structural support when its wall is removed to make openings. Heesoo Kwak has undergone many trials and errors as he began to open up the side of the tube in his works. The southern volume on the second floor is open on all the sides to ensure a view towards the courtyard to the north and the adjacent park to the south. To complement this, only one row of three columns is arranged to form a Vierendeel structure. There is no column at all below the floor, just like a bridge. The structure of the tube is completely dismantled here; nonetheless, the continuity of Il-pil-seo is still alive.

From Design to Behavior
In his previous works, Heesoo Kwak seems to have been interested in the formative aspect of architecture itself. All formative attempts of line, diagonals, cantilevers and unity, were made to gain a dynamic of escape from everyday life. Architecture can't fly: however, the dynamics of forms established a way to imagine that architecture could move or fly. His numerous experiments and attempts gave him a lot of practical experience about the influence of a specific architectural language on people's behavior. Now in Gardenus, his interest seems to have shifted towards how to prompt visitors' behavior through architecture. On the slope are placed four low wooden benches, two inside and two outside. Along a gently descending slope, a spatial device on which people can lie down ensures its own area and front view as well as provides comfort, both physically and mentally.
Unusual dynamic forms created with oblique lines and slopes include the potential risk of ending up as merely visual vanity, unless accompanied by appropriate programs. Among the basic elements of architectural form such as walls, floors, and ceilings, the floor has a direct influence on the behavior of people, as it is the mechanism that transfers loads and controls our movement and sight, having direct contact with our body. In so many examples

三层混凝土凉亭一景
view from the concrete pavilion on the third floor

运动和视线,并且与人们的身体有直接的接触。在世界各地的许多建筑案例中,建筑的台阶设计被认为是唯一要用到斜面元素的设计。庭院中低矮的木质长椅也可以被理解为台阶的变体,但它的功能不仅限于单纯地供人们歇脚和观景,还可以供人们进行更多样、更自由的活动。

主体部分和辅助部分

郭熙秀在先前的作品中,会把一些辅助构件暴露出来,以保证主体空间的纯粹性,如:像前锋一样的柱子支撑浮在上方的管状结构,满足最起码的承重要求。这些辅助构件在数量设置上十分严格,作为必要元素,它们显示了受力的走向,保证了空间的通透和主次结构的分明。然而,在坡州庭院中,有的辅助构件开始逐渐被隐藏起来,在东南角添加的楼梯就是一个例子。附设这个楼梯的初衷是为了在结构方面支撑二层的主体部分,在功能和流通方面不是必需的存在。换句话说,很难分清楚它属于主体部分还是辅助部分。在这里,郭熙秀似乎做了两种选择:第一,清晰地展现其组成结构,也就是明确区分主体结构和辅助结构;第二,消除或隐藏这些结构,而模糊主体和辅助构件的关系,追求构图上的通透性,曾是现代建筑的重要常态之一。考虑到建筑语言在这个作品中得到了很好的运用,我们不禁要产生这样的疑问:当代建筑设计是否仍旨在展示其组成结构?

太阳马上就要落山,人们还在庭院里享受着时光。三层的混凝土凉亭只有三面,分别由屋檐、地板和右手边的墙体组成,向西面和北面开放。这个空间也像是一个敞开的院子。我坐在台阶上低矮的木质长椅上,往后靠靠,就能看到天边的晚霞罩在微微隆起的黛青色山脊上,美丽得令人陶醉。

throughout the world, tiers are regarded as the only program planned for the sloping floor. The low wooden benches in the Gardenus can also be interpreted as a variation of the tiers, but they allow more diverse and freer activities besides the simple function of sitting and watching.

Main and Ancillary Parts

In his previous works, Kwak has exposed additional elements in order to maintain the purity in the main space, for example, a prop-like column that supports the floating tube from below. This column was the minimum requirement to withstand the weight of the tube. These ancillary elements reveal the flow of power as necessary, neither more nor less, ensuring transparency of structure and the definite hierarchy of main and ancillary parts.

Ancillary elements in Gardenus are gradually hidden. The stairs added to the southeast corner exemplifies the fact. The role of staircase is given to this volume, originally added to support the main volume floating on the second floor. In terms of function and circulation, the stairs are not that necessary: in other words, it became difficult to distinguish the ancillary element from the main element. Considering that the architectural language and its use are properly applied to this work, this raises a question: Kwak seems to have had two choices – first, a clear exposure of the composition, that is, to clearly distinguish between the main and the ancillary; second, to erase the composition or to make it invisible. In other words, to make no distinction between the main and the ancillary, while transparency of composition was one of the important norms of Modern architecture. But it leaves the question as to whether contemporary architectural design still aims to reveal its composition.

Soon, the sun began to set: people were still enjoying their time around Gardenus. The concrete pavilion on the third floor is framed on only three sides by eaves, floor, and a wall on the right-hand side – it is open to the west and the north. This space too seems like an open courtyard. Sitting back on the low wooden bench on the tiers, I could enjoy the beauty of a red sunset over the low bluish ridges of the horizon.

相聚

Come T

人是社会动物，需要一些建筑把我们聚到一起。这也解释了为什么世界上先有公共建筑，后有家庭建筑。如今，公共建筑的类形和建筑设计方法五花八门。有的建筑师从他们在社区出生成长的经历中找到动力，想要在设计上反映自己社区的风土人情。

本项调查研究在建筑案例的选取上，不去考虑当代大众化的公共场所，而是聚焦一些功能

We need buildings that bring people together, because we are a social species. Communal constructions may even predate domestic architecture. Nowadays the typology of communal buildings and the approaches to their architecture are very diverse. Being rooted in a community gives architects good incentive to reflect the vernacular, but not all do. In this survey, the buildings we sample ignore the most popular contemporary places of

卡索拉山的"高清"医院_High Resolution Hospital Center in Cazorla / EDDEA
安德马特音乐厅_Andermatt Concert Hall / Studio Seilern Architects
克雷斯波洛图书馆_Kressbronn Library / Steimle Architekten
阿尔科塔农民起义100周年纪念馆_Memorial and Monument to the 100th Anniversary of the Alcorta Farmers Revolt / [eCV] estudio Claudio Vekstein_Opera Publica
MO现代美术馆_MO Modern Art Museum / Studio Libeskind

相聚_Come Together / Herbert Wright

上更老派、更基于社区的建筑。这些建筑主要见于小型社区,在历史上最近的城市爆炸式发展之前,人们都是聚居在这样的小社区当中的。即使在这种时光静止、相对保守的居住环境中,我们依然可以发现这些建筑同样传达出了现代生活的美好与活力。

gathering, which are shopping centers, transport hubs and dance clubs. We consider several buildings where the function is older and more community-based. They are mainly sited in smaller communities, where most of humanity has been based until the historically-recent explosion of urbanisation. In this more conservative and timeless settlement environment, we nevertheless find that contemporary expression in architecture is alive and well.

相聚
Come Together

Herbert Wright

人类必然以群居的方式生活，进化的要求使得人类必须形成组群，个体在群体中的生存机会比独自生存要高。然后，组群变成共同生活的部落。甚至在城市产生之前，就已经出现了具有公共用途的建筑。在土耳其的哥贝克力出土了可能有11000年历史的巨石圈，它们显示出一种社会和文化上的功能。也就是说，我们最古老的建筑是社区建筑。

我们主要考察了一些小镇的社区建筑。那里的社区比大城市的社区联系得更紧密，所以我们只举一个城市社区的例子。我们期望可以看到地方优势在建筑设计中发挥作用，比如，使用当地的材料和传统来定义建筑的地方色彩。这与把不知从哪来的建筑直接"空投"到一个地方的做法恰好对立，后者完全不顾建筑所处的地方环境，而当地的现实情况并非如此简单……

高清医院（2018年）位于西班牙安达卢西亚的内陆区，由塞维利亚的EDDEA建筑师事务所设计（第86页）。"高清"（西班牙语：Alta Resolución）这个名字听起来多少有些独特，它在这里不是指显像技术，而是指在医疗管理上"要高效地清除疾病"。它建在卡索拉山脉的一个山坡上，与附近一个约有7600人的小城隔河相望。建筑师采用当地的一种土褐色砖块制作了建筑立面，把一些大型的凹窗错落有致地嵌在了墙面内，还用火山砾石铺了屋顶。经过这些设计，这座狭长而低矮的直线形建筑把自己融进了周围崎岖、干燥的丘陵地貌当中。从小镇那个方向看，医院如同从橄榄树丛中长出来的二层建筑，前面有些单层的翼房，仿佛消失在线性上层的平面轮廓之下。其中一些翼房还拐出直角，围出了一个旱地植物园，将建筑与耐旱的灌木和松树混合在一起。因为植物园不是全封闭的，所以不能称其为院子。建筑师对自然景观的运用并没有就此止步，而是将其引入建筑内部，直接借景造景。从上方看，一层的屋顶有许多的圆孔，这些圆孔其实是开放的天窗，通过曲面玻璃墙贯通到室内地面，与室内其他空间隔离，再把室外的植物引入到玻璃墙中间。一层室内花园走向上与二层的平行。

Humans need to gather together, because evolution hard-wired us to form groups, in which the individual has a higher chance of surviving than alone. Groups became tribes with a shared communal life. Even before they settled in urban communities, built structures serving a communal purpose emerged. Circles of stone megaliths, perhaps eleven millennia old, have been excavated at Göbekli Tepe in Turkey. They suggest a social, cultural function. Our oldest known architecture is a community building.

We survey community buildings mainly in small towns. There, the local community is more coherent than in the big city, from which we take just one example. We may expect local strengths to play in the designs, such as the use of local material and traditions in style that define vernacular architecture. That's the opposite of architecture air-dropped from elsewhere that doesn't care where it is. And reality on the ground is not so straightforward..

The High Resolution Hospital (2018) by Seville-based architects EDDEA (p.86) is in inland in southern Spain's Andalusia. It takes it name from its approach to health care management rather than any particular technology. Built on sloping ground, it faces the adjacent town of Carzola (population 7,600) across the valley of the small river. The facades are earthy brown local brick, punctuated by large, recessed windows and the roofs are covered by volcanic gravel stone. That helps the long, low rectilinear building blend into the rugged, dry and hilly landscape. From the town, the hospital appears amongst the olive groves as a two-storey volume, its one-storey wings lost under the flat profile of the linear upper storey. Some of these wings turn at right angles, embracing arid garden. This mixes the building into the landscape of dry scrub vegetation and forest pines. These gardens are not courtyards, because they are not entirely enclosed. But the architects have gone further with natural landscape, bringing it inside the building itself. Seen from above, the ground floor roof is

土耳其的哥贝克力石阵
Göbekli Tepe in Turkey

医院另一侧是路面结实的矩形停车区,车辆就在那里进出。医院一端有个悬挑式的圆弧形雨篷,下面就是医院入口。医院内部雪白的墙面散发着未来主义的氛围,而褐色的天花板又抵消了医院那种冰冷的感觉。内部的采光也十分不错,阳光通过落地观景窗或者玻璃观景墙进入室内。顶楼设有病房和诊室。医院后侧是大楼的另一条长边,它笔直的立面正对高速公路下边的山坡。尽管如此,山坡被切割开,建筑被石笼墙挡住了,从这两个方向伸出了一条通道,倾斜向下到达较低的楼层。

这种整体上与周围地形保持紧密联系的设计,让人想起了另一个西班牙建筑杰作——由涅托-索韦哈诺建筑师事务所打造的麦地那扎哈拉博物馆(2009年),整个建筑从未超出地面一个楼层的高度。它在空间构图上也采用了矩形,设有多个通往地下的路口,与医院的不同主要体现在材料和户外空间上。像麦地那扎哈拉博物馆一样,高清医院成为西班牙建筑中与景观完美融合的另一个经典案例。

接下来,我们选择了Seilern建筑师事务所设计的安德马特音乐厅(2019年,第104页)。瑞士著名的安德马特滑雪度假区被迷人的群山环绕,那里常住居民不足1500人。然而,音乐厅周边的环境却没有专属于高原街道的那种古雅气息(度假区被建在了公路边上)。相反,它身后赫然耸立着新建的丽笙酒店。丽笙酒店是面向国际市场新开发的项目之一,酒店大楼高7层,体型颇长,对音乐厅来说是一个庞然大物。长久以来,瑞士的一些度假区一直很受游客欢迎,但气候变暖导致雪线上升,未来无法继续依赖滑雪这一板块,于是度假区决定开发新的旅游项目,音乐厅就是其中之一。除了能丰富旅游项目,建造音乐厅还提供了一个契机,让度假区可以借此重新找回它的魅力与它的地方特质,这些东西也在以和滑雪场一样甚至更快的速度消失。幸运的是,音乐厅的选址工作十分顺利,有个原本用来建造会议中心的场地,由于工程中止,留下了一

penetrated by a number of circles, which are open skylights for a stretch of vegetation separated from interior common areas by a full-height curved glass wall. This internal garden runs parallel to the upper storey. Another wing defines a hard surfaced rectangular bay which vehicles can access, with islands of vegetation. At one end of the building, a cantilevered curving canopy marks the entrance. The interiors are futuristic and white, but their warm brown ceilings dispel a clinical appearance. Illumination is generous, either by the full-height windows with landscape views, or the filtered light coming through the internal garden. The upper storey is specialised service and hospital rooms. The back of the hospital is the other long side. It is a straight facade facing a slope climbing to a highway. Despite this, the hillside is cut into and held back by a gabion wall, and an access road slopes down from both directions to a lower-level floor.

The overall close relationship with the terrain is reminiscent of another outstanding Spanish building, the Madinat al Zahara Museum (2009) by Nieto Sobejano Arquitectos, which never rises more than a storey above ground. It is also a composition of rectangular spaces and is accessed by paths descending into the ground. However, the materials and open spaces are different. Like the museum, the High Resolution Hospital is another example of Spanish architecture that brilliantly melds with its landscape.

Our next project is the Andermatt Concert Hall (2019) by Studio Seilern Architects (p.104). The Swiss popular ski resort of Andermatt (with less than 1,500 permanent inhabitants) is surrounded by stunning mountains. However, the immediate surroundings are not quaint, distinctively Alpine streets – they lie on the other side of a highway. Instead, long seven-storey facades of a recently-built Radisson Blu hotel, part of a new cluster of developments for the international market, loom over the concert hall. Swiss resorts have long attracted visitors, but can no longer count on snow for their future, and diversification is a response to the rise in altitude

欧洲新会议中心，福克萨斯建筑工作室设计，意大利罗马，2016年
EUR Congress Center by Studio Fuksas, Rome, Italy, 2016

个2000m³的地下混凝土箱体，给音乐厅提供了现成的结构。

位于伦敦的克里斯蒂娜·塞勒恩的建筑工作室以惊人的设计理念传达了建筑魅力及其地方特质的必要性。废弃的地下体量向上挤出身子，拔地而起，长成了一个崭新华美的音乐厅。厅外的公共区域做了景观改造，人们透过音乐厅一侧的玻璃墙可以看到厅内容纳663人的礼堂。礼堂上方吊着三台声音反射装置，玻璃墙在保证采光的同时，又能让外部处于地面位置的人能看到其中两台声音反射装置。屋顶板以一定的斜角向外悬挑出来，轮廓十分醒目，下方的声音反射装置看上去就像微微张开、刚好把里面的珍珠露出来的蚌壳。音乐厅传声效果的设计是与比利时的卡勒音效公司联合完成的。这些声音反射装置宛如夜晚发光的石雕，极大地增添了建筑的视觉诱惑。这种设计估计只能找到一个先例，在视觉效果上更具规模，这就是由福克萨斯建筑工作室设计、位于罗马的欧洲新会议中心，它的"云朵"设计让整座建筑同音乐厅一样具有类似的视觉效果，室内的悬浮构件如同洁白而神秘的云朵漂浮在玻璃幕墙的后面。

礼堂内部从斜天花板到下面的深色地板都采用木材装饰。与之形成对比的是一些金属感十足、由三角形板构成的光滑墙面，墙板正对着栏杆，位于比它体型更小、近乎环状的天窗展馆下面。整个音乐厅从内到外都透出了它的精巧别致，安德马特地区也因此多了一座令人耳目一新的特色建筑。有人觉得这个音乐厅不属于真正的地方社区，但在否定它之前，我们必须要回溯那些古代场所的社区精神，例如，英国的巨石阵，它不也一样吸引了来自世界各地的人吗？但它依然保有地方特质，这个来自4000年前的范例很好地证明了这一点。

谷仓是用作储存农产品的建筑。标准的谷仓建筑至少要追溯到13世纪，当时英格兰的科吉歇尔庄园建了一座36m长的谷仓并保留至今，它那倾斜的屋顶及下面方盒状的木质框架已成为谷仓的标准形式。谷仓一直未在社区建筑的名单

of snow-lines due to climate change. A concert hall offers diversification, plus an opportunity to bring some beauty and identity to a place where they are melting away even faster than the snow. Luckily, the site of the concert hall provided a pre-existing structure for it - a 2,000 m³ concrete box sunk into the ground for an uncompleted convention center.

The London-based practice of Christina Seilern delivers the necessary beauty and identity with a striking architectural idea. The abandoned underground volume is extruded vertically into a new pavilion above ground level, which is glazed around one side, so that the 663-capacity auditorium can be looked into from new landscaped public realm outside. That brings natural light in, and also gives ground level visibility to two of three floating acoustic reflectors. They are suspended from a bold roof which is angled and cantilevered out, as if an oyster shell had been opened just enough to reveal its treasure. The acoustics were a collaboration with Belgian company Kahle Acoustics. These reflectors are like sculpted stones which light up at night. Their visibility makes them a deeply seductive architectural device, with perhaps just one precedent. On a bigger scale, The Cloud within the EUR Congress Center, Rome (2016) by Studio Fuksas, has a similar effect - a mysterious white abstraction floating behind a glass screen.

The auditorium interior is lined with wood, from the ceiling's angled planes to the dark floor below. That contrasts with the full-height metallic mirror triangles that make up the wall opposite the bar, which is situated under a smaller, almost circular skylight pavilion. Inside and out, this concert hall is an ingenious, refreshing addition to Andermatt. And before dismissing it as not really local, we should remember that ancient community sites such as Stonehenge, England, also drew people from far and wide - and it was built 4,000 years ago.

A barn is an architectural typology to store agricultural produce. At least since the 13th century, when the

科吉歇尔庄园谷仓，英格兰
Coggeshall Grange Barn, England, UK

中，即使最近几十年住宅改建浪潮席卷了北美和西欧，这个情况也依然没有改变。直到最近，在德国南部一个叫作克雷斯波洛的、人口为8600的村子中，谷仓作为一种社区资源得到了重生。克雷斯波洛图书馆（第116页）由德国斯图加特市的Steimle Architekten建筑师事务所设计，保留了谷仓原有的外壳，建筑师将巨大的瓦片屋顶悬挑到建筑体量之外，一层上方的木制框架扩建结构把屋顶悬挑的部分支撑起来。一层用混凝土结构进行了翻新，里面开放的大厅可以用来举行各种活动或展览。图书馆的主要楼层在大厅上方，露出了其周围的木框架结构。主层上面还悬坐着一个更窄小的画廊，它正好卡在斜屋顶开始向上收缩的高度，阁楼平面和木质桁架在这一层开始合拢。一层混凝土基座上边的实墙被垂直木板外部屏障包围，建筑师还把木板旋成不同的转角，这样它们就不会创造出一个平坦的表面了。光线透过竖直的玻璃窗透入屋内。建筑师实际上就是通过改旧屋来造新屋的，甚至让木框架上的金属紧固件就那么露出来，所以整座建筑看起来还有一些现代气息。尽管如此，建筑师通过对原始外围护结构的采用，对传统谷仓木立面的重新演绎，成功地完成了向地方语言和地方特质的献礼。

阿根廷圣达圣非州以西300km有一个叫作阿尔科塔的小镇，人口仅有7450人。1912年，一场农民起义以这个小镇为发源地，向全国开始蔓延。1962年，小镇为了纪念这次起义中的英雄事迹和斗争精神建了一些雕塑，以写实的风格表现了当时的苏维埃社会主义思想。

然而，先前的这个纪念工程只是在远离镇中心的边缘地带上铺几个混凝土地基，安上几尊塑像罢了。于是，当地又在这片地带建造了纪念这场起义100周年的展馆（第132页），建筑师克劳迪奥·维克斯坦另辟蹊径，颠覆了常规的纪念性建筑的设计。它的造型十分抽象，走向呈直线型，内部的结构被外显出来，周围的景观完全开放，与这些特点相似的恐怕只

surviving 36m-long Coggeshall Grange Barn in England was built, a timber-framed rectangular box under a pitched roof has been the standard form. A barn is not a community building and that's still the case after the surge of residential conversions that has swept North America and Western Europe in recent decades. But in the south German village of Kressbronn (population 8,600), a barn has been reborn as a community resource. The Kressbronn Library (p.116) designed by Stuttgart-based Steimle Architekten retains the barn's original envelope, with a large tiled roof overhang beyond the building volume supported by extensions of the wooden structural frame above the ground floor. The new ground floor is concrete and includes an open area for different programming such as events or exhibitions. The main library floor is above it, revealing the wooden frame around it. It is overlooked by a narrower gallery floor which floats almost at the height where the pitched roof starts to narrow the open attic space and its spans of timber trusses. The solid walls above the concrete base of the ground floor are surrounded by an exterior screen of vertical timber planks, rotated so they don't create a flat surface. They filter light through vertical windows. This conversion is effectively a new building (even the timber frame shows its metallic fastenings) and it looks contemporary. Nevertheless, with its original envelope and clever re-interpretation of wooden barn facades, it salutes the vernacular.

In Argentinian town of Alcorta (population 7,450), 300km west of Buenos Aires in the state of Santa Fe, a revolt by agricultural workers started in 1912, and spread across the country. In 1962, the struggle and heroism of this event was to be expressed with statues in the style of Soviet socialist realism, but the project only got as far as concrete foundations on the edge of town. The Memorial Space and Monument to the 100th Anniversary of the Alcorta Farmers Revolt (p.132) is built on these, and the architect Claudio Vekstein has taken a very different architectonic approach which entirely rejects that of normal monuments. Its abstract form, linear extent, the exposure of its structure, and its wide-open landscape context, are probably comparable only with

政治迫害受害者纪念馆,彼得·卒姆托和路易丝·布尔乔亚设计,挪威瓦尔德,2011年
Steilneset Memorial by Peter Zumthor with Louise Bourgeois, Vardø, Norway, 2011

有建筑师彼得·卒姆托在挪威瓦尔德设计的政治迫害受害者纪念馆(2011年)了。然而维克斯坦的纪念馆在构图上颇具解构主义的色彩。场馆的立面饰有棱角分明的金属纹理,内部是一个可容纳150人的礼堂和展览馆。一面巨大的屏障把整个场馆遮蔽,它折叠起来悬挑出了建筑体量。屏障外侧的材质由树脂和玻璃纤维板构成,土褐色的外观和粗糙的质感,既是对土地的隐喻,也是对工人们的象征,象征着他们身上的皮肤,象征着他们可敬的精神。沿着横穿屏障的斜坡就能来到屋顶上的观景台,这里是一个由混凝土和波纹金属构成的独立展馆,钢结构也在上面升起和折叠,但只留下了一个裸露的框架。

红色是国际公认的象征着英雄主义的颜色,在这座纪念馆中也得到了广泛的使用。建筑一侧有一个柏油路面的人民广场、人工草坪和一块新造的绿地。丰富的材料和科技感十足的建筑风格混搭,使整个纪念馆的构成令人叹为观止。这个位于乡间的历史纪念馆几乎在每个游客的心里都留下了难以磨灭的印象。

我们现在把视角从小镇的社区建筑转向时尚精致的都会建筑——一座新建的私家美术馆。和刚才提过的安德马特音乐厅一样,这座美术馆也是为来自世界各国的参观者服务的。MO现代美术馆(第148页)位于立陶宛首都维尔纽斯(人口54.4万),场地的前身建筑是立陶宛电影院(1965—2005年)。2014年,全球解构主义建筑大师丹尼尔·里伯斯金的工作室在建筑竞赛中拿到了这个场地的设计权。这个项目对派乐默大街的旧广场进行了改建,只是从地面面积来看,新广场的规模并没有之前大,但是把楼上各层的空间加起来,那广场面积就非常可观了。美术馆声称要以派乐默大街为界,在中世纪老城区和周边城市之间建立一条走廊,正因如此才需要建设多层的广场。整座建筑呈一个巨大的白色方块,设计师在方块的一角"切出"了一个豁口,从这里引入一条通往内部的街道。这是一种解构主义的做法,因为他真的把建筑体量

the Steilneset Memorial (2011) to the victims of witch-hunts in Vardø, Norway by Peter Zumthor with Louise Bourgeois. However, Vekstein's memorial is a deconstructivist composition. A corrugated metal volume of angular planes houses a 150-capacity auditorium and a gallery, and is masked by a vast panelled screen which folds to overhang this volume. Its outward side is a matrix of resin and fibreglass panels which are brown and rough, like the soil and the skin of the workers who are honoured here. A ramp across the screen leads to the viewing platform on the roof of a separate pavilion of concrete and corrugated metal, and steelwork rises and folds over it too, but is left as a bare frame.

The colour International Red, associated with heroism, is used extensively. Beside these structures is a hard-surfaced civic plaza, landscaped grass, and a newly planted patch of woodland. The whole composition, its mix of materialities and expression in high-tech style, is extraordinary. Few visitors would forget this rural monument.

We now move on from small-town community buildings to look at a sophisticated city's new private art museum. Like the Andermatt concert hall, it serves a non-local cosmopolitan audience. The MO Modern Art Museum (p.148) in the Lithuanian capital, Vilnius (population 544,000) replaces the Lietuva cinema, built in 1965 but closed in 2005. Studio Libeskind, a global giant of deconstructivist architecture, won the architectural competition for the site in 2014. The project offers a new public plaza, although it builds over the cinema's previous plaza on Pylimo Street, which was larger at ground level. But there is more plaza at higher levels, and the museum's claim to create a gateway between the medieval Old Town and the surrounding city, which Pylimo Street divides, is connected to that. The white incised rectangular block is entered from the street by a big angular cut in one corner, a gesture suited to deconstructivism because it literally deconstructs the volume.

解剖了。从这个建筑切口进入街道的确让人兴奋不已，这种设计可以追溯到1977年位于美国加州首府萨克拉门托的BEST商店的切口型建筑。它的设计师詹姆斯·瓦恩斯和他的事务所SITE做出了一种视觉错效，让建筑的边角看上去好像被拆离了，等商店关门后又重新被装上去。如今，这座位于维尔纽斯的建筑也失去了一个边角，创造了一个5m长的悬挑结构，形成了一个通往美术馆的入口。这个缺角同时还作为一个开放空间的起点，向上贯通整座建筑，形成一条走廊。从入口对面的角落，可以看到一个抬升的广场平台，楼梯高8.4m。三角形空间和嵌装玻璃、锐化的边角和倾斜的墙壁，将这两对组合混搭在一起，这是里伯斯金典型的设计风格，这些元素在这座美术馆中得到了充分的体现。建筑中唯一在几何上呈现弯曲的元素就是螺旋式的二层楼梯，黑色的扶手和周围白色的空间形成了对比。在构图上，设计师将其中的极简主义风格本身进行了解构，可谓匠心独运。

　　尽管大城市之外保守主义尚且存在，但是在我们所调查的小镇建筑当中，能保留当地乡土风格的例子屈指可数，当代的建筑思想已在小镇建筑中非常活跃，复兴这样的小镇社区很重要，建筑就是社区复兴的要素之一。至于大城市，位于维尔纽斯新建的MO美术馆取代了曾经在全国具有重要意义的电影院，这既是文化上的大势所趋，也是城市更新的必然结果。添加了新的用途之后，一座激动人心的绅士化建筑就此诞生，地区绅士化就是以一群对流行趋势十分敏感的都市精英取代当地原来的社区人群的现象。但是，无论对于保守的地方性社区还是世界性社区，它们都有了支持社区建筑建设一个新的深层次原因：社区建筑中的群体互动完全不同于数字世界中新型的互动模式。社交网络、游戏网站、约会网站、线上论坛和流媒体公司等——它们都举办社区活动，但不是实地参与和身体接触的方式。现实世界中的社交生活才是至关重要的，而虚拟世界正在侵蚀它，如果我们真的意识到这点，就应该不断地更新我们的社区建筑。

The drama of entering through a sheared-off corner can be traced back to the BEST Store's Notch Building (1977) in Sacramento, California, where James Wines' architectural practice SITE created the illusion that the corner had been torn away and separated, although it was moved back into place after business hours. In Vilnius, the missing corner creates a 5m cantilever and provides access to the museum, It is also the start of an open void rising through the building, which becomes a gateway. Stairs rise 8.4m to a raised plaza terrace cut into the corner three-dimensionally opposite the entrance. The building itself is a characteristically Libeskind mix of triangular spaces and glazing, sharp angles and sloping walls. The only curved element is a spiral staircase rising two storeys, its black balustrades a counterpoint to the white space around it. The composition, in which minimalism itself is deconstructed, is masterful.

In the small town buildings we survey, there is little of the vernacular surviving, but contemporary architectural ideas are alive, despite the conservatism found beyond big cities. It is important to re-invigorate such communities, and exciting community architecture has a role to play. As for great cities, the new Vilnius art museum, replacing a cinema that once had national importance, is about cultural mega-trends and urban renewal. An exciting architecture with a new use fits in well with gentrification, which replaces a city's local communities with a trend-sensitive metropolitan elite. But whether a community is local and rooted, or cosmopolitan, there is a new and profound reason to support community architecture. A community building is a place in which group interaction is totally different to the new interactions of the digital world. Social networks, networked gaming, dating sites, online forums, streamed events etc. – they all host community activities, but without location or physical contact. If we recognise that social life in the physical world is vital, and that the virtual world is eroding it, we should constantly renew our community buildings.

卡索拉山的"高清"医院
High Resolution Hospital Center in Cazorla
EDDEA

西班牙哈恩省的乡村地区多为低矮的山地地貌和丘陵地貌，形成了一个天然的圆形剧场，发源于卡索拉山脉的瓜达尔基维尔河及其各个支流都流经此地。几个世纪以来，这里一直都是最能清晰反映出安达卢西亚地貌特点的代表性地区，也是古罗马文化遗产的一部分。我们在这里的全景图中找到了自己所处的位置，也就是卡索拉山脉自然公园的入口。

建筑与景观

医疗基础设施的建设通常对其周围环境会产生很大的影响。该医院大楼紧邻自然公园，靠近城镇，凭借着地理位置的优势，让环境在病人的护理治疗中发挥积极的作用。

建筑场地面积为30000m²，场地地形复杂多变，为了避免建筑对周围景观的破坏，建筑师采用了和谐统一的设计主题。

医院中心在结构上并没有采取线性设计，因此也没有长长的、平整的建筑立面。相反，建筑物以很自然的姿态在整个地块铺展开来，与周围环境融为一体，以至于人们从远处很难辨认出它的存在。不过这么做并没有影响医院的功能效率。

建筑分为三层。主入口位于中间层，这一层的公共概念最为突出，具有开放的功能和多样化的空间关系。其特色在于它的各个入口和公共空间的节奏，如广场和庭院，为满足每一种特定的功能需求，其空间的定向都经过了最优化的选择。体量最小的是顶层，只限于提供病房和手术室等操作性的功能空间，而底层区域则实现了让马路附近的外部空间同这个原本受限空间的连通。

由于挡土墙和入口道路的建设，建筑的外围区域受到了一定的改动，建筑师在这些区域种植了针叶林、灌木丛以及一些森林植物，对其进行修复。这些植被的存在增强了建筑的美感，使其与周围特殊的地理环境更为和谐融洽，保留了人们对这个场所的记忆。

结构和材料

哈恩省是陶土之乡，陶瓷制品，尤其是砖瓷，是当地常见的材料。而如今，建筑师将它们用作现代建筑材料，通过现代烧制工艺实现了对通风立面的需求。

建筑师在选择材料的过程中，最感兴趣的就是材料是否具有一种特性，能够激发或改变人们对空间或环境的感知。所以，虽然这座小城的建筑普遍都是白色的立面，但这座医院却不是，建筑师想让它和这个地块、这个乡间，以及卡索拉自然公园融为一体。为此，建筑师不仅使用了瓷砖墙面，还用火山砾石铺了屋顶，这些材料采用了相同的色系，将建筑包裹起来。此外，从任何角度看去，植被的栽种都有助于减少项目带来的视觉影响。

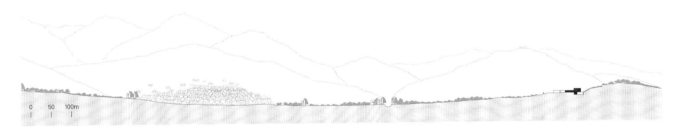
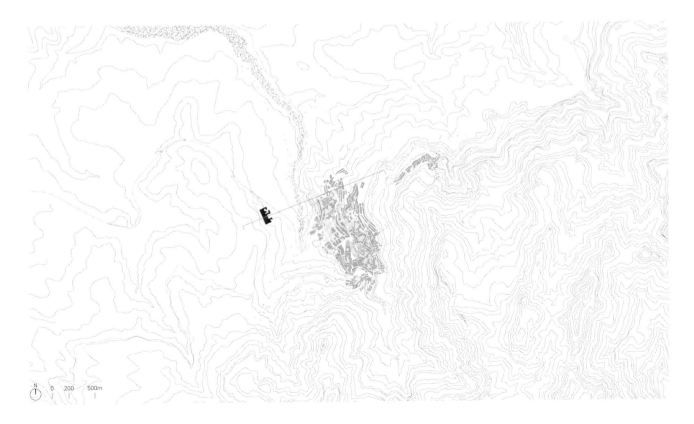

建筑立面的抽象化设计表现了医疗机构应该具备的每一个必要特征。由于施工缝数量多得惊人，尺寸也很夸张，因此立面上出现了大大小小的孔径组合，呈现一种在秩序与凌乱之间摇摆不定的特性。在建筑内部，建筑师让素净洁白的隔墙同天然的石头和通透的玻璃和天窗并置在一起，让它们相得益彰。

Jaen countryside is characterized by low ridges and hills where the land creates a natural amphitheater crossed by the Guadalquivir River and its tributaries. For centuries, the region has represented the clearest example of the Andalusian landscape. Being part of the Roman cultural heritage trail, this panorama guides us to our location, on the threshold of Sierra de Cazorla Natural Park.

Architecture and Landscape

Generally, medical infrastructures have a very strong impact on their surroundings. Taking advantage of the location, right next to a natural park and close to the town, the architecture emphasizes the relationship between patient and environment as a step towards therapeutic care.

On a plot of 30,000m², with varied topography, the architects avoided impacting on the scenery by taking the theme of unity into consideration.

The Hospital Center is not a linear construction, so does not have a long and plane facade; rather the building spreads through the territory and merges with the environment in a natural way, avoiding being recognized from the distance. This does not affect its functional efficiency.

项目名称：High Resolution Hospital Center in Cazorla
地点：Carretera A-319 Cazorla-Peal de Becerro, Cazorla, Jaén, Spain
建筑师：EDDEA – Ignacio Laguillo (project and supervision works site),
Luis Ybarra (project and supervision works site), Harald Schönegger (competition and project)
施工地点：EDDEA – Eladio Suarez (quantity surveyor), Roberto Alés (quantity surveyor)
发起人：Andalusian Regional Government, Andalusian Health Service
合作方：Project and supervision works site – Miguel Sibón (industrial engineer),
Enrique Cabrera (civil engineer); Supervision Works Site – Ignacio Olivares (architect),
Blanca Farrerons (architect), Javier Salvador (architect), Jaime Fernández (architect),
Alejandro de la Torre (quantity surveyor), Jaime García (landscape designer, V.Olimpia);
Labor Risk Prevention_Project and supervision works site – Jesús Martínez (Prevencoor);
Competition and project – José María Sánchez (architect), Carlos Serrano (architect)
承包商：DRAGADOS – Antonio Vecino (civil engineer, delegate);
Francisco J. Luque (quantity surveyor, group chief);
Hilario Nuñez, Carlos Cabana, Pascual Molina (quantity surveyor, works site in chief)
造价：EUR 17.811.341,34 / 建筑面积：9,610m²
竞赛时间：2007
设计时间：2007—2017
施工时间：2009—2018
摄影师：©Fernando Alda

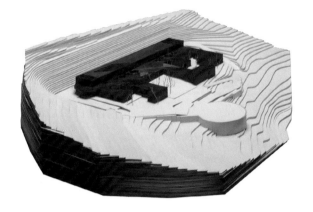

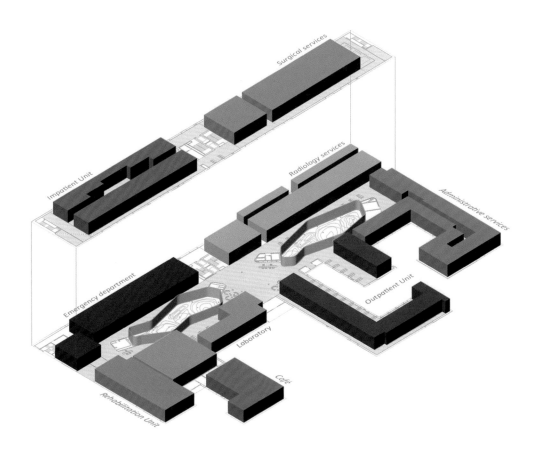

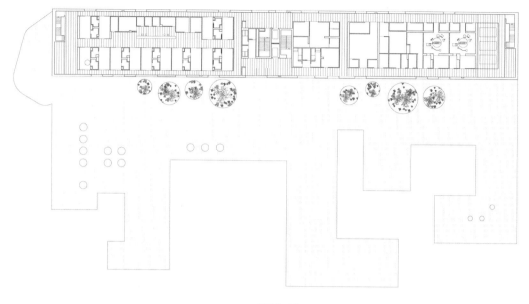

二层 first floor

一层 ground floor

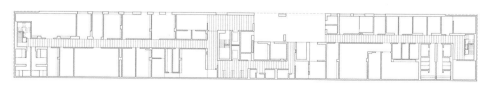

地下一层 first floor below ground

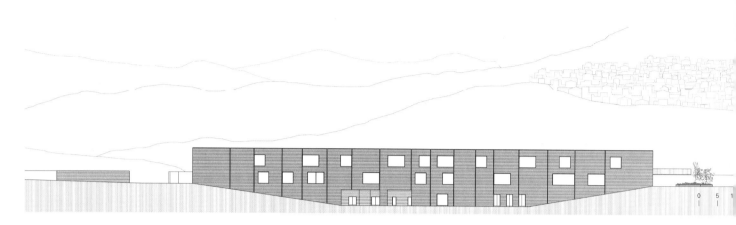
西南立面 south-west elevation

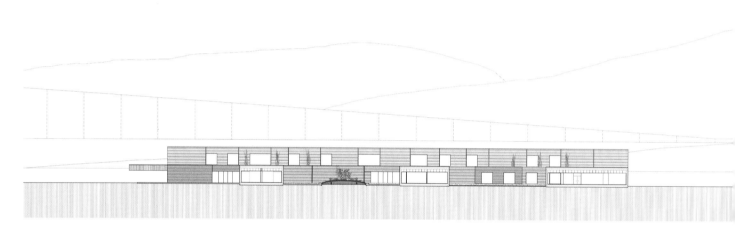
A-A' 剖面图 section A-A'

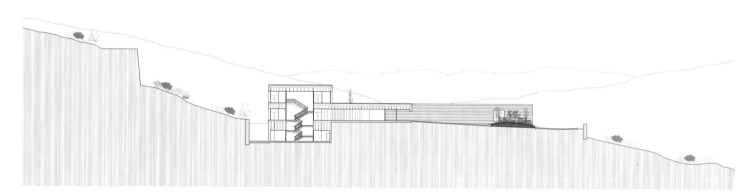
B-B' 剖面图 section B-B'

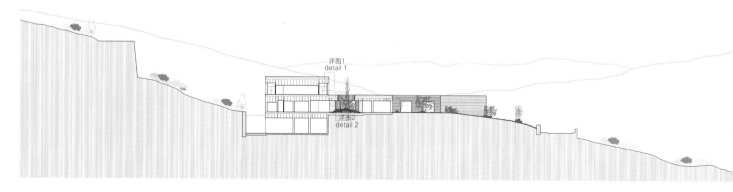
C-C' 剖面图 section C-C'

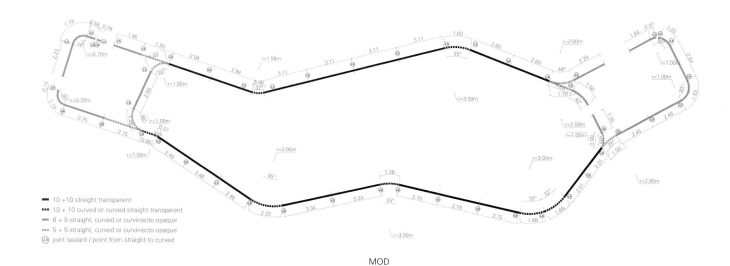

- 10 +10 straight transparent
- 10 + 10 curved or curved straight transparent
- 8 + 8 straight, curved or curvirrecto opaque
- 5 + 5 straight, curved or curvirrecto opaque
- joint sealant / point from straight to curved

MOD

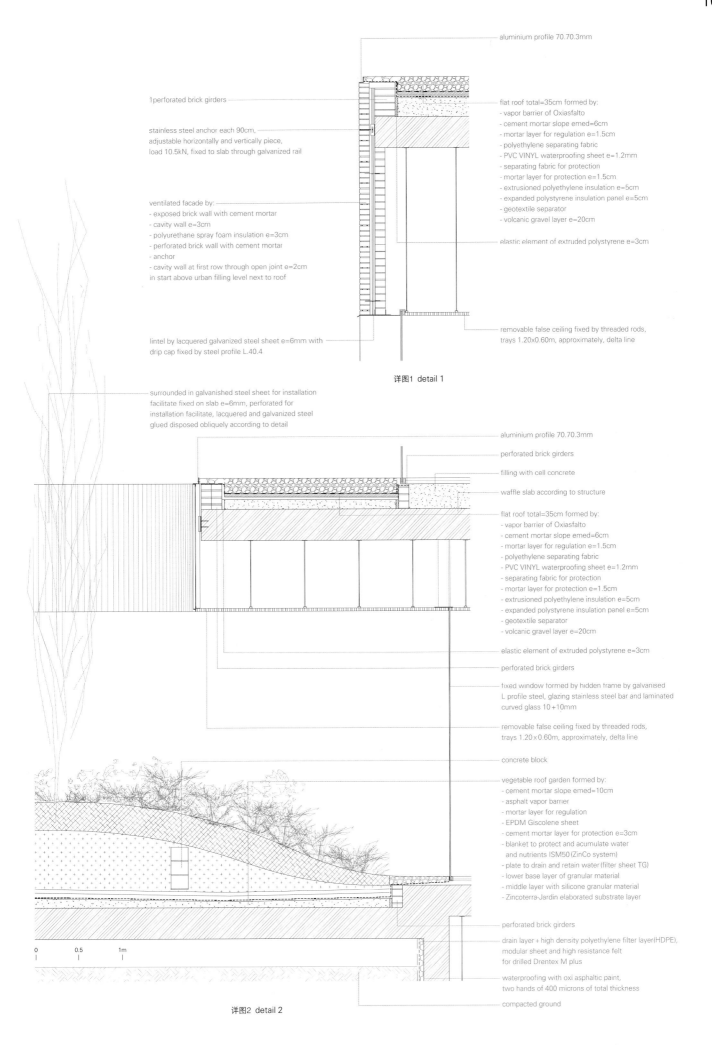

详图1 detail 1

详图2 detail 2

The building is distributed across three floors with the main entrance on the intermediate floor, the most public floor which holds open functions and diverse relations. This level is distinguished by its accesses and the rhythm of public spaces such as squares and courtyards that are positioned at the optimum orientation in order to satisfy each particular function. The upper floor is the smallest and its program is limited to operational functions such as wards and surgeries, while the below ground area keeps the otherwise restricted spaces accessible from the outside by a road.

Conifers, bushes and forest plants recuperate the exterior spaces that have been modified by the need for retaining walls and access. This vegetation contributes to the aesthetic, of a building integrated into its exceptional location, preserving the memory of the place.

Construction and Materials

Jaen is a land of pottery, and ceramics – especially bricks – are a common local material. However, today they are used as modern construction materials, made with modern techniques and fulfilling the requirements for a ventilated facade.

The most interesting characteristic for the architects in the process of selecting the material was its ability to inspire or transform the perception of the space or the environment. Instead of using white, the predominant color in the town, the architects wanted the hospital to blend in with the territory, the countryside and the Sierra de Cazorla Natural Park. This was achieved not only with the use of ceramic, but also with the volcanic gravel used for the roof, and which surrounds the building in the same color spectrum.

In addition, the vegetation helps to reduce the visual impact of the project from any point of view.

The abstraction of the facades represents each necessary feature that such an institution demands. The construction joints appear in an exaggerated manner, both in number and size. This creates a composition with different apertures in the facade, an uncertain quality between order and chaos.

Inside the building, the neutrality of the white partitions coexists with the natural stone and transparency of the glass and skylights.

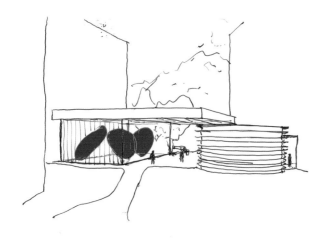 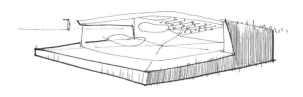

安德马特音乐厅
Andermatt Concert Hall

Studio Seilern Architects

该项目由瑞士安德马特阿尔卑斯有限公司和比利时建筑巨头BESIX集团联合开发,由Studio Seilern Architects (SSA) 建筑师事务所负责设计,成为首个为阿尔卑斯滑雪度假区特别打造的音乐殿堂。

安德马特是一个位于瑞士腹地的自治山区,该地区一直在努力转型为一个世界级的四季度假区。音乐厅所在位置坐拥安德马特的全景,它的建成标志着该地区的转型臻于成熟。场地的功能区包括一个新建的广场,音乐厅位于广场中心,附近是新建的酒店和一些木屋。

该音乐厅属于一个改造项目,其场地原本要用来建造一些会议中心和酒店,计划废止后,留下了一个有效体积约为2000m³的地下混凝土箱体。

SSA建筑师事务所提议将现有屋顶的一大部分抬高,这样大厅的有效音量就会扩大一倍,空间体积也会增加,可以容纳75人的交响乐队、663人的观众席。

将屋顶抬高以后,设计师也有了更多的创作空间。他们借此机会让音乐厅看起来宛如度假村的一尊雕塑,同时对传统音乐厅在空间上的封闭性和内向性也重新进行了思考。通过采用玻璃立面,音乐厅充盈着自然的光线。冬天来的观众会看到自己被窗外的飞雪围绕,夏天来的观众会感到自己被大自然和阳光所拥抱,这便是设计师们创造出的浪漫。

从街道往里看,厅内的声音反射装置像是漂浮在空中的公共艺术品。街上的行人可以看到音乐厅,从街上观看里面的观众和乐队表演。SSA建筑师事务所希望音乐厅能够表现出步行街活力奔放的一面,而不是像一般意义的音乐厅那样,如一个捂得严严实实的箱子。

大厅可能被用于举行管弦乐表演、摇滚音乐会、会议等各种活动,因此需要根据要求灵活地改变座位布局。为此,设计师采用了一种可伸缩的系统,可以让多达9排的阶梯式平台隐藏在二层主厅的下面。只需要几分钟,这个圆形的立体式剧院就一改它的私密性,变成了一个12m高的开放性空间,接受自然光线的洗礼。

音乐厅的声学设计极为考究,从根本上保证了人声和器乐声可以被清晰地传递给观众,让观众感觉仿佛置身于音乐的怀抱。为了实现这种效果,音乐厅内部设计的形态和布局得到了优化,把从舞台传来的声音反射到观众席的各个角落。楼座倾斜的前部和雕塑感十足的木质天花板等各种平面如同从地面上升起的波浪,将整个空间包裹起来,使观众仿佛置身于音乐的浪潮中,其视觉体验和听觉体验通过厅内的几何构造均得到了很好的提升。

当观众透过天花板眺望远处阿尔卑斯的山景时,它们可以看到上方悬挂的声音反射装置,它们如同雕塑一样,给整个视野增添了活力。本身从上方和外部观看舞台就是一种独特的体验,而这些"雕塑"的存在更是让这种独特感锦上添花。

被抬高和扩建的屋顶形成了一个有顶的广场,也为音乐厅提供了另一个独立的入口,来宾可以从邻近的酒店直接进入。

原来的混凝土楼梯间也被改造为一个温馨的门厅。门厅的倾斜墙壁采用切面反光玻璃包覆,该灵感来自阿尔卑斯山冰川和岩层的表面。

Designed by Studio Seilern Architects (SSA) and realized by Andermatt Swiss Alps and BESIX (Belgium), the Andermatt Concert Hall is the first major purpose-built concert hall in any Alpine ski village.

The music venue is located in the panoramic area of Andermatt, a mountain village and municipality in the heart of Switzerland. The hall represents the culmination of the area's transformation into one of the world's finest year-round destinations. Facilities include a new village square, at the heart of which is the new concert hall, and where new hotels and chalet facilities are also located.

The project transforms an existing underground space – a concrete box with an effective volume of approximately 2,000m³ – originally intended for conventions and hotel events.

SSA's proposal was to raise a large section of the existing roof, doubling the effective acoustic volume, and increasing the total capacity in order to host a full 75-piece symphony orchestra and 663 audience seats.

Raising the roof gave SSA the opportunity to create a sculptural object within the village and rethink the traditional notion of a concert hall as a closed and inward-looking space. Thanks to a glass facade, the concert hall is awash with natural light. The romantic idea was that during a winter concert, the audience would be surrounded by a whirlwind of snow, and in summer, by nature and sunshine.

From street level, the acoustic reflectors are seen floating over an empty space, like a public art piece. Passersby can see into the concert hall, witnessing the audience and orchestra from the street as a spectacle. SSA wanted the concert act to become an active frontage to the pedestrian street, rather than a closed-off box that this building typology offers.

The hall is used for different purposes, and requires the flexibility to host different seating layouts – from an orchestral performance to a rock concert or a conference. This flexibility is achieved via a retractable system that allows up to nine rows in a stepped platform to disappear under the main balcony. In a few minutes, an intimate theatre in the round is transformed into a 12m-high open space, washed by natural light.

Excellent acoustics are essential for speech intelligibility, musical clarity and a sense of being enveloped by music. These are provided by the optimized interior topography of the hall to reflect the sound from the stage to every part of the audience. Surfaces, such as the inclined balcony fronts and the sculptural timber ceiling, wrap the space, giving the audience the impression of being inside a wave of music which is visually and acoustically enhanced by the interior geometry of the hall.

A sculptural acoustic reflector suspended from the ceiling contributes to the dynamism of the view – which looks out across the Alps – rendering the unusual experience of looking into the performing space from above and outside, even more intriguing.

The raised and extended roof creates a covered plaza and an alternative entrance to the concert hall, independently accessible from the adjacent hotels.

The existing concrete staircase core are also transformed into a welcoming foyer, whose inclined walls are cladded in faceted reflective glass, inspired by the surfaces of the glaciers and rock formations of the Alps.

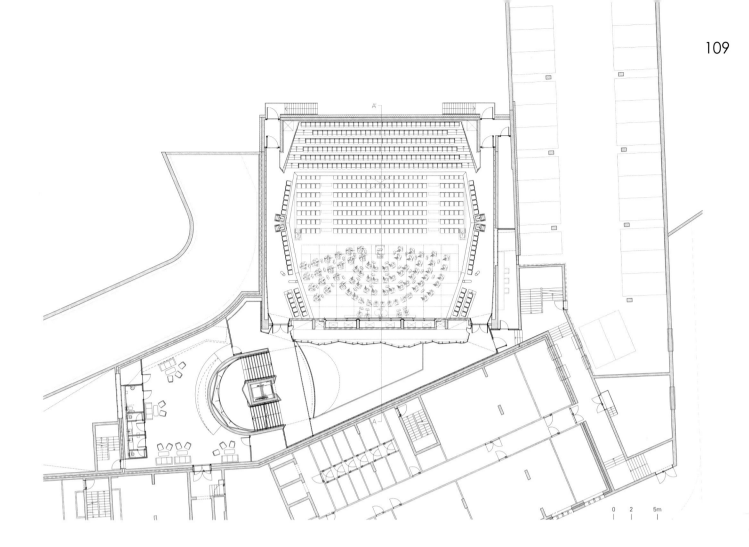

项目名称：Andermatt Concert Hall / 地点：Andermatt, Switzerland / 建筑师：Studio Seilern Architects / 项目团队：Christina Seilern, Marcos Velasco, Alberto Favaro, Brigitta Hadju, Jonathan Wrynne, Ruby Law, Tasos Theodorakakis, Hana Potisk, Sonia Theodosiadi, Enrique Pujana / 声学顾问：Kahle Acoustics
剧院顾问：Ducks Sceno / 结构工程：Suisseplan / 机电工程：BESIX / 景观设计：Hager / 防火顾问：AFC / 照明设计：Michaeljosefheusi GmbH / 承包商：BESIX
客户：Andermatt Swiss Alps & BESIX / 总建筑面积：2,072m² / 造价：11m CHF / 设计时间：2017 / 施工期间：2018—2019
摄影师：©Roland Halbe(courtesy of the architect)-p.105lower, p.106~107, p.108, p.112~113, p.114, p.115 ©KanipakPhotography (courtesy of the architect)-p.110

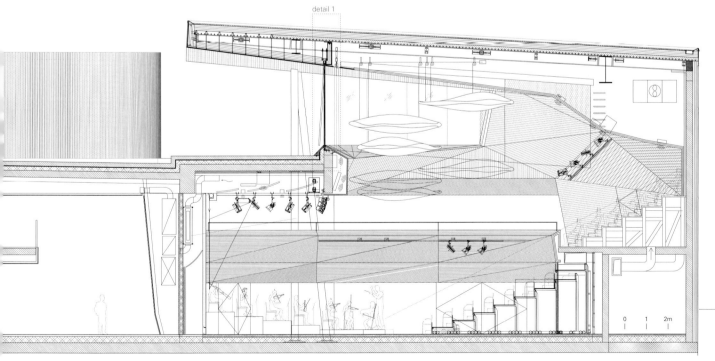

A-A' 剖面图 section A-A'

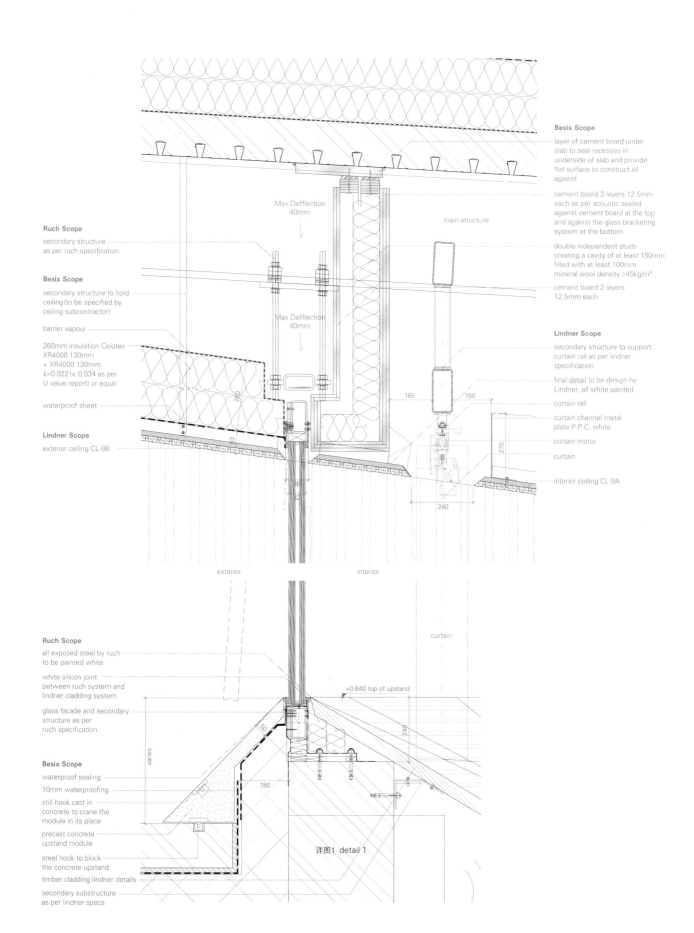

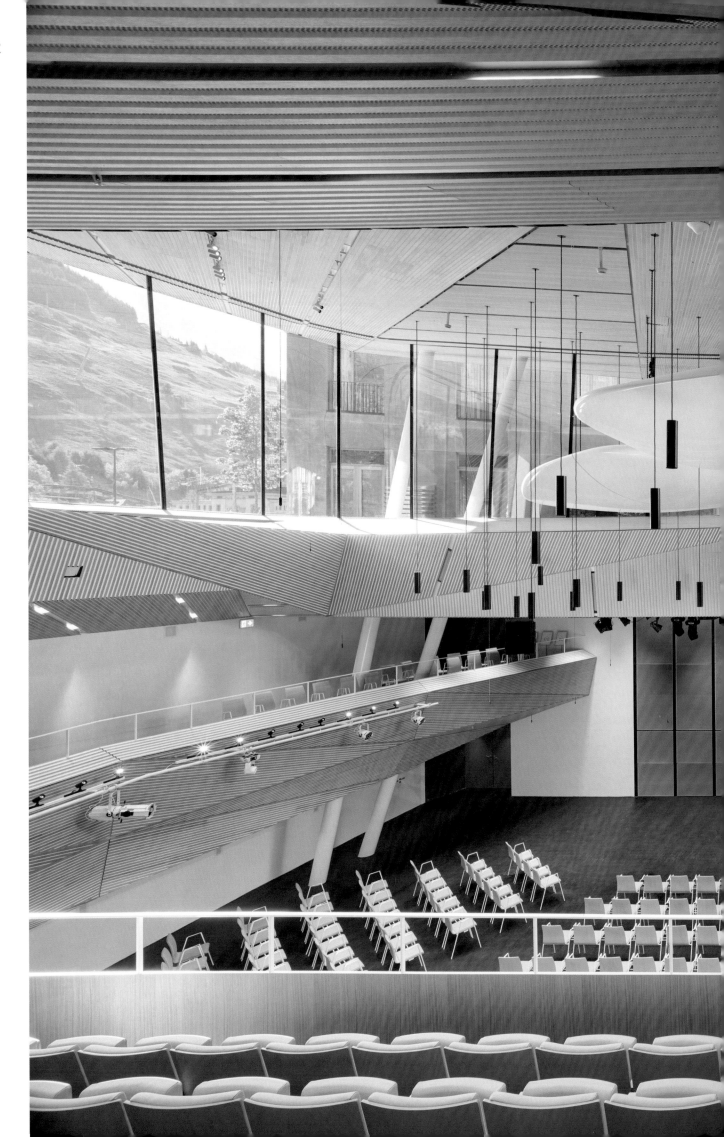

克雷斯波洛图书馆
Kressbronn Library

Steimle Architekten

克雷斯波洛图书馆设计的指导思想是建立在尊重建筑遗产的基础上,因此设计既要保留当地旧有谷仓的特点,还要将其改造成一座现代化的、出入便捷的建筑。

改造工作要进行周密考虑,设计不能破坏农家建筑的建筑特点,既保留了原有的深长的悬挑屋顶,也保留了一层为加固层、二层为打谷场的传统分隔模式。

精致的木质镂空装饰取代了旧的立面外观,这些装饰非常照顾原有立面给人的感受,对其小心翼翼地进行修改。考虑到谷仓建筑的历史意义和人文情怀,设计师把这种风格移植到了现在的这座图书馆上,不需要多少建筑手法,也不需要去压制它的历史表达。

该谷仓位于村庄的显著位置,又靠近镇公所和节日大厅,因此,它被用作图书馆和社区中心。具有了全新的公共用途之后,它就成了一个重要的节点,将三个独立的节点(图书馆、镇公所、节日大厅)连接到了一起。

引人驻足的前院和宽阔的室外露台彰显了谷仓的新公共特性,鼓励人们在此逗留。本来是一座内向型的仓储建筑,如今成了一座开放的建筑,外观上保持着熟悉的形象;同时,所使用的材料和一层偌大的门窗也清楚地表达出了它的现代风格。

谷仓外观的改造简单明了,却都经过深思熟虑。谷仓的新地基采用了构造均匀的保温混凝土结构,门窗宽阔大方,保留了石材坚实、深沉的传统形象。建筑中的新洞口不仅给室内带来了更多的日光,还通过底层大型的玻璃门窗把室内与室外连接在一起,带给人们一种全新的体验方式。建筑的核心部分仿佛从混凝土基座中生长出来,连接着室内的二层和三层,开放的走廊创造了一种明显与众不同的空间体验。尽管如此,马鞍形状的屋顶显得十分熟悉亲切,从屋顶伸到墙面之外的木桁架看起来历史悠久,这种传统建筑的式样依然没变,从许多方面都能观察得到。

开放的前厅表现出了对游客的欢迎姿态,并鼓励游客之间的交流和接触。一层有多种用途:包括一个可分隔的多功能厅、一个展厅和一个24小时开放的图书馆。图书馆里有媒体库、杂志库,还设有阅览区,让整栋建筑的视野惊人地开阔。新老结构在这里进行了一场激动人心的对话。新旧结构之间的平衡成了该建筑的特质。

为了维持旧屋顶的内部特征,建筑师在对其修复时态度十分谨慎,仅在必要时才更换单个木质部件。屋顶下方的护墙板原先围绕着打谷场,现在成为修复工作的重要参考依据。

为了获得图书馆所需要的漫射光线,设计师将木板竖直对齐,并沿着垂直轴旋转,悄悄地改变了原来的立面形式。这些细长的木板条形成了木栅,可以过滤阳光,使其渗透到内部,而其优雅的造型给建筑外观平添了一抹现代色彩,令人眼前一亮。

The guiding principle for the design of Kressbronn Library was to treat the architectural heritage with consideration and respect, preserving the character of the old barn while transforming it – with just a few, well-considered interventions – into a modern, accessible building.

The conversion of the former agricultural building should preserve its character, retain its deeply overhanging roof and also maintain the traditional division between the solid ground floor and the threshing floor above.

Delicate open woodwork now replaces the old facade, reacting to the existing structure sensitively and gently altering it. Given the historical significance of the stable's architecture and the emotional ties people have to it, the building was resituated in the present, using few architectural means and without suppressing its history.

At its prominent location within the village, close to both the town hall and the festival hall, the barn's new public use as a library and community center has established an important connection that now links the three separate buildings.

An inviting forecourt and a broad outdoor terrace reinforce the barn's new public character and encourage people to linger. What had been a rather introverted storage building has now become an open house that conserves its familiar image while at the same time speaking an unequivocally modern language

1923

2018

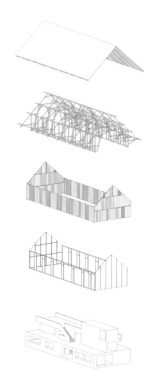

我一直是现在的我,但现在的我不同于过去的我。
I was always what I am now, and yet now I'm different from what I was.

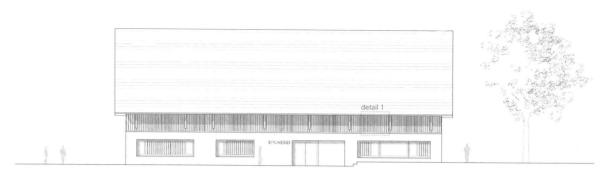

东立面 east elevation

南立面 south elevation

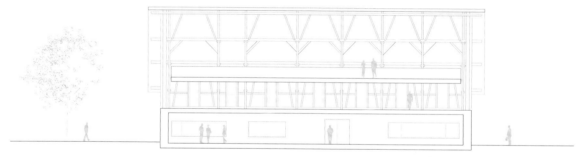

A-A' 剖面图 section A-A'

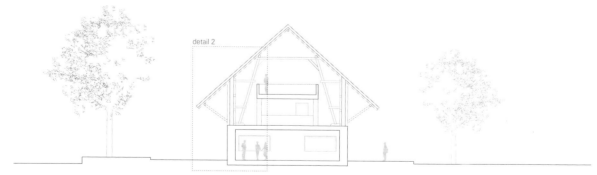

B-B' 剖面图 section B-B'

项目名称：Kressbronn Library / 地点：88079 Kressbronn a. B., Germany / 建筑师：Steimle Architekten BDA-Thomas Steimle / 客户：Commune of Kressbronn a. B. / 总建筑面积：860m² / 体积：3,500m³ / 施工方：massiv, timberwork 立面：fairfaced concrete, timber formwork / 竣工时间：2018.8 / 摄影师：©Brigida González

三层——走廊 second floor_gallery

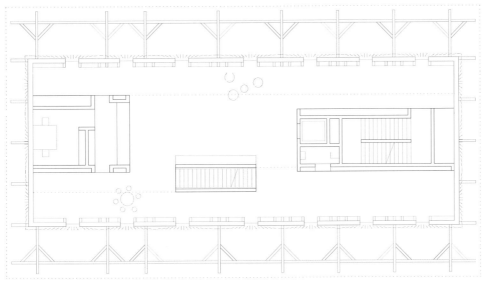

二层 first floor

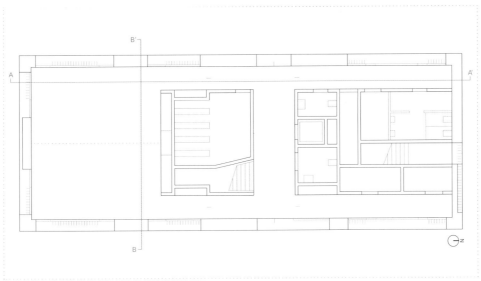

一层 ground floor

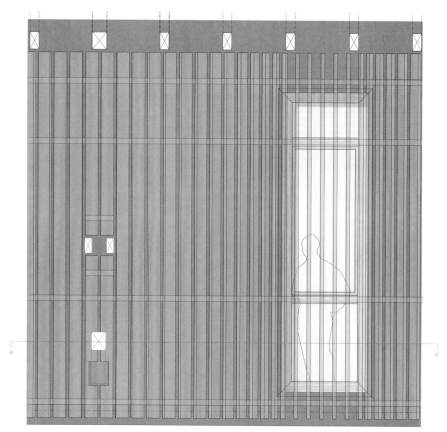

详图1 detail 1

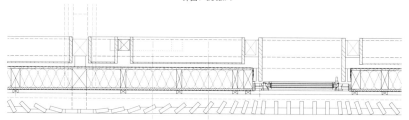

a-a' 剖面详图 detail a-a'

with the materials used and the large openings on the ground floor.

The barn was revamped with simple, precisely conceived interventions. Its new base, homogeneously constructed of insulating concrete with deep, generously proportioned reveals, preserves the solid impression of the original stone floor. The new openings not only admit significantly more daylight, but also enable the interior to be experienced in an entirely fresh way – inside and outside open up to each other on the ground floor through the large glazed openings.

The core emerges from the concrete base, connecting the two upper floors within, and its open gallery creates a tangibly different spatial experience. Nonetheless, the familiar saddleback roof, with its historical timber trussing that projects out far beyond the facade, remains present in the space – and can now be viewed from many perspectives.

An open lobby welcomes visitors and encourages communication and encounters. The ground floor can be put to a variety of uses: as a dividable multipurpose room, an exhibition space, and a 24-hour library. The library on the first floor, with its media and magazine gallery and its reading stations, offers surprisingly open views throughout the building. Here in particular, the old and new enter into an exciting dialogue. The balance of past and present becomes the building's special quality.

In order to maintain the internal character of the protective roof, the old roof structure was carefully restored, replacing individual wooden parts only where necessary. Beneath the roof, the typical clapboard siding that originally surrounded the threshing floor is the important reference.

To obtain the diffused light desired for the library, the old facade was subtly transformed by aligning the wood boards upright and rotating them along the vertical axis. This results in slender wooden slats whose varied angles filter daylight, allowing it to penetrate inside – and whose elegant structure bestows an astonishingly modern appearance to the building.

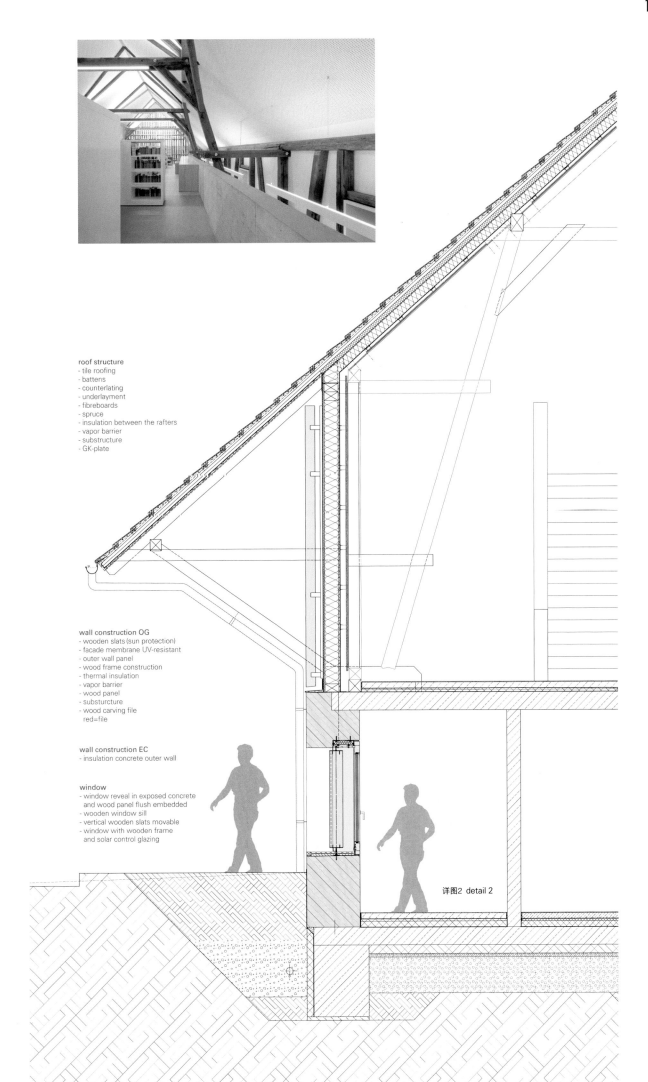

roof structure
- tile roofing
- battens
- counterlating
- underlayment
- fibreboards
- spruce
- insulation between the rafters
- vapor barrier
- substructure
- GK-plate

wall construction OG
- wooden slats (sun protection)
- facade membrane UV-resistant
- outer wall panel
- wood frame construction
- thermal insulation
- vapor barrier
- wood panel
- substurcture
- wood carving file
  red=file

wall construction EC
- insulation concrete outer wall

window
- window reveal in exposed concrete and wood panel flush embedded
- wooden window sill
- vertical wooden slats movable
- window with wooden frame and solar control glazing

详图2 detail 2

阿尔科塔农民起义100周年纪念馆
Memorial Space and Monument
to the 100th Anniversary of the Alcorta Farmers Revolt

[eCV] estudio Claudio Vekstein_Opera Publica

这座纪念馆是为了歌颂是名为"阿尔科塔之歌"的农民土地革命而建的,革命者大部分是意大利和西班牙移民。革命从阿尔科塔镇中心开始,然后扩大到整个圣菲省,最后遍及全国,最终成立了阿根廷农业联盟(FAA)。

项目组与FAA委员会通力合作,得到了各级政府的大力支持,让这个纪念馆成功地唤起那段革命记忆,缅怀了那些革命农民,他们团结一致、不屈不挠,坚决维护自己对土地的使用权和所有权。不仅如此,还让这些记忆真实地呈现在农民及其他居民的日常集会场所中,削减了纪念建筑以往常有的、令人敬而远之的肃穆感,使其更加平易近人。

这个长100m、宽75m的不毛之地,原本只伫立着四个裸露的大型混凝土地基。这些地基要追溯到1962年,时值土地革命50周年,该镇为此建了一些纪念性雕塑,并使其呈现出雕塑家薇拉·穆欣娜的苏维埃社会主义写实派的风格,这些地基就是该工程的一部分。如今的这座纪念馆就是在FAA的委托下开发的,它占地400m²,设有可承办多种小型文化活动的功能区,包括一座用于公众举办庆祝或纪念活动的广场、一个能容纳150人的礼堂、一个能举办常设或临时展览的画廊,里面陈列着从各省收集到的1912年革命留下来的历史文物。

室内各个空间联系紧密,而室外则由西北侧的大面积折叠外立面覆盖,包括具有律动感的巨型钢结构门廊、斜框架和凹凸不平的模块化屏墙。这个屏墙设计得十分经典,它巨大的尺寸使其成为90号公路上一道亮丽的风景,吸引了沿途旅客和小镇居民的目光。

巨型屏墙的浮雕式纹理使人联想起在FAA的照片档案中看到的那些仓库的麻袋和装在玉米拖车上的口袋。设计师选择了当地的镶板工艺,把经过玻璃纤维强化的树脂作为面板,将粗麻布在木质几何像素模具上压模,才做出了具有这种浮雕式纹理的板材。

西侧展馆是FAA的办公室和公共卫生间,其结构形式采用了钢混骨架,外围是拔地而起的钢结构挡墙。在西侧展馆的上方就是一个下倾的平台,平台一侧连着屏墙外侧的坡道,另一侧连着楼梯。从平台望去,远处广阔的地平线和南美草原壮美的落日,引起了人们的无限遐想。

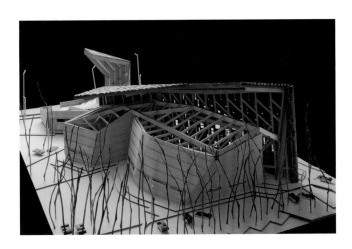

设计师让材料呈现出粗糙程度不同的质地，形成一种织物般触感的纹理。通过这种建筑语言，表现出农民在田间耕种的辛劳，这一道道纹理就像他们被烈日灼黑的皮肤，布满了如车辙般的皱痕，还像他们身上裹着的粗衣麻布和身后堆起的粮食麻袋。艺术家安东尼奥·贝尔尼在他1934年的画作《游行》中用的就是这种表现手法，侧面烘托了阿尔科塔游行时的喧闹。建筑的立面被重建为自由堆叠的墙体，各个门廊在水平方向上从墙体中凸出来，共同形成了连续的截面。带有凹槽的表面在建筑之外展开，而金属板的横向切片最终保证了建筑物与地平线协调一致。

建筑师对草甸进行了修整，将潘帕斯草原的风光引入室内。以胶合板材料包裹室内空间，最大限度地给展厅争取了可用的空间。室内沟壑状板材，连绵向上的凹槽和小麦色的弯折构造淋漓尽致地展现了南美平原广阔无垠的风貌。

This monument celebrates the 1912 agrarian revolt, led by a small group of mostly Italian and Spanish immigrant tenant-farmers known as "El Grito de Alcorta". With the town of Alcorta as its epicenter, the rebellion spread throughout the Santa Fe province, and later the whole country, giving rise to the Argentine Agrarian Federation (FAA).

Working with the FAA Assembly and collaborating with the Alcorta Commune, the Santa Fe Province and the Federal Government, the memorial evokes the farmers, their work and struggles, their use and possession of the land and their co-operativism; it also actualizes them in a daily gathering space for farmers and citizens, overcoming the reverent passivity of monuments of the past.

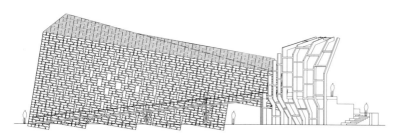

西北立面 north-west elevation

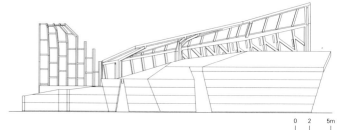

东南立面 south-east elevation

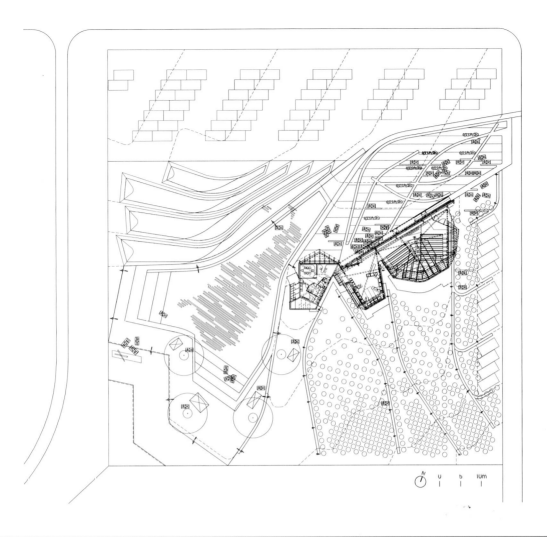

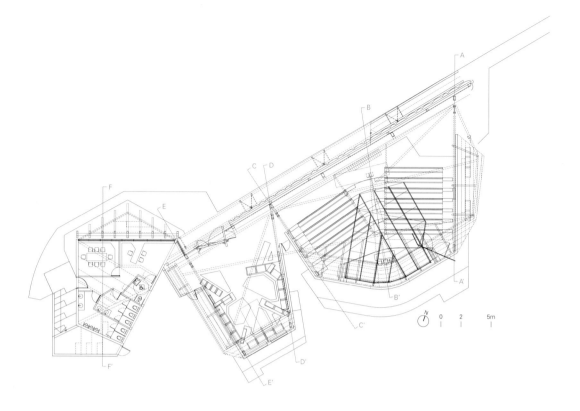

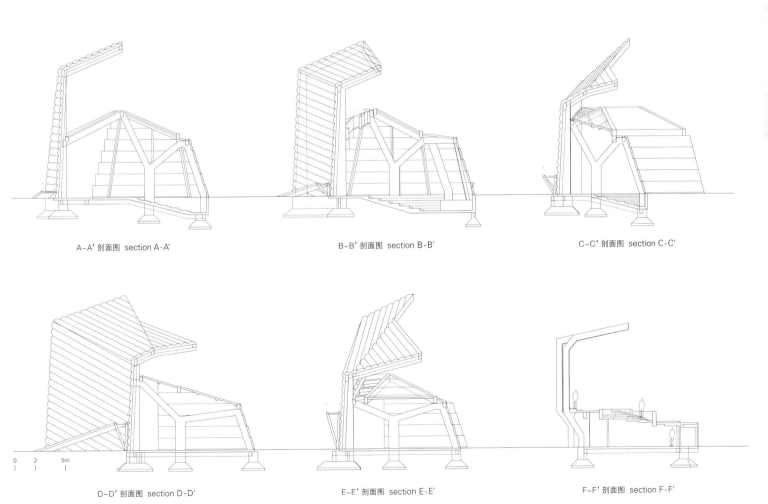

A-A' 剖面图 section A-A'   B-B' 剖面图 section B-B'   C-C' 剖面图 section C-C'

D-D' 剖面图 section D-D'   E-E' 剖面图 section E-E'   F-F' 剖面图 section F-F'

Only four of the large, exposed concrete foundations remain in the bare 100m x 75m terrain; these were built in 1962 for the 50th anniversary of the uprising as part of a monumental project of sculptural figures in the style of Vera Mukhina's soviet "socialist realism". The current 400m² building program developed with the FAA fulfils several small-scale cultural functions: a civic plaza that allows the realization of public acts of celebration and remembrance, an auditorium for 150 people, and a gallery of permanent and temporary exhibitions: an interpretation center displaying historical artefacts from the 1912 events, brought together from the provinces.

The interior's intense and intimate scales are housed by an extensive folded exterior plane on the northwest side, structured by large rhythmic steel porticos, inclined frames and rugged modular panels. This screen articulates classic scenery with its monumental scale, creating a forced perspective for those arriving on the tracks circulating on Route 90 or from the town. The massive relief recalls stockpiles of burlap or tow corn sacks, based on salvaged images from the FAA photo archives; it materializes through a system of panels crafted locally out of resin reinforced with fiberglass and crude burlap molded on geometrically pixelated wooden forms.

The west pavilion contains FAA offices and public conveniences, in a reinforced concrete structure with steel profiles emerging from the plowed earth. Above it, on the descending terraces – accessed by an exterior ramp along the main screen and a staircase following the stepping – the vast horizon and splendid Pampean sunsets can be contemplated.

Rough textures at different scales create a textile and tactile grain. This evokes, through the language of architecture, the

labor which the agrarian workers engraved on the land; their tanned skins, rugged as furrows cracked by the sun, their clothes and their rough bags stashes in piles – just as artist Antonio Berni observed in his 1934 painting "Manifestación", which alludes to the Grito de Alcorta. This facade is reconstructed as a free stacking wall with horizontally extruding porticos that conjugate to form continuous sections. The fluted surface unfolds beyond the building, while the lateral sectioning of the sheet metal panels finally reconciles the building with the horizon.

The exterior plowed pampas are invited into the interior. Here, plywood surfaces extend to achieve the maximum possible space for exhibitions, wrapping the auditorium in furrowed boards, ascending grooves and wheatears.

项目名称：Memorial Space and Monument to the 100th Anniversary of the Alcorta Farmers Revolt / 地点：Route 90km 78, Commune of Alcorta, Santa Fe Province, Argentina / 建筑师：[eCV] estudio Claudio Vekstein_Opera Publica
主持建筑师：Claudio Vekstein / 项目经理：Carolina Telo / 项目助理：Mariana Pons, Pedro Magnasco, Mercedes Peralta, Martin Flugelman, Santiago Tolosa, Stephen Wanderer, Alisha Rompre, Elizabeth Menta, Pamela Galan, Dolores Cremonini, Maca Cerquera, Shaghayegh Vaseghi / 效果图：Hernán Landolfo
景观设计顾问：Elena Rocchi, Lucia Schiappapietra, Teresa Rozados; Assistants: Cecilia Chiesa, Clara Miguens / 结构顾问：Tomás del Carril, Javier Fazio / 板材顾问：Mark West, Ronnie Araya
照明设计顾问：Giuliana Nieva / 施工管理：Province Department of Architecture and Engineering (DIPAI), Special Projects Unit, Ministry of Public Works and Housing, Santa Fe Province / 承包商：Coirini S.A.
结构承包商：Héctor Malo / 客户：Argentine Agrarian Federation, Government of Santa Fe Province, Commune of Alcorta, Government of the Argentine Republic
用地面积：7,500m² / 建筑面积：400m² / 设计开始时间：2011
竣工时间：2018 / 摄影师：©Federico Cairoli (courtesy of the architect) - p.132~133, p.137middle, p.139, p.141, p.144, p.145; ©Sergio Gustavo Esmoris (courtesy of the architect) - p.134, p.135, p.136, p.137bottom, p.140, p.142, p.143, p.146~147

展厅平面图 exhibition room plan

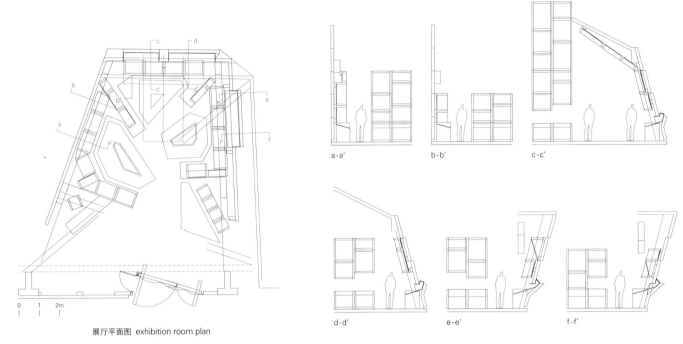
展厅剖面图和家具立面
exhibition room sections and furniture elevations

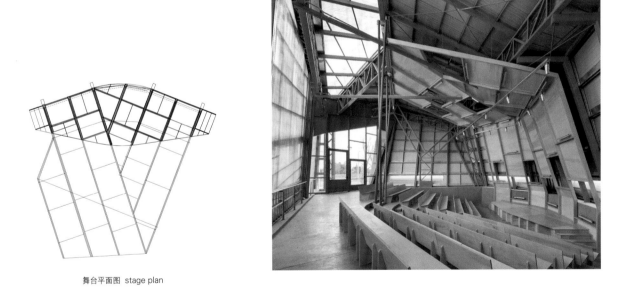
舞台平面图 stage plan

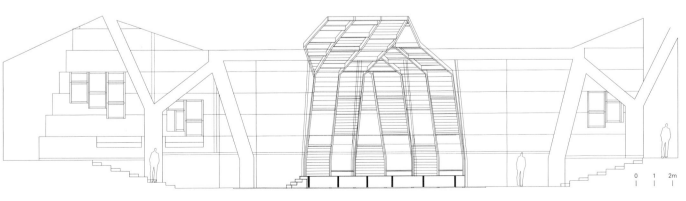
舞台展开立面
stage unfolded elevation

MO 现代美术馆
MO Modern Art Museum

Studio Libeskind

MO现代美术馆的馆长米尔达·伊万诺斯基恩说："这座美术馆是立陶宛和维尔纽斯市的文化里程碑。这个新的世界级美术馆将展示本土艺术，并将探索它们与全球艺术界的联系。同时，那些从未面世的艺术品也将通过这个场馆展示给全世界。"

这座占地3100m²的美术馆是对维尔纽斯市的过去和现在的重新诠释，它整合出了一个新的公共广场，距离历史闻名的中世纪古城区仅几步之遥。MO美术馆被视为连接两个经典时代的文化"门廊"，一侧是18世纪的近代城市格局，另一侧是被城墙围绕的中世纪城市。其灵感源自于这座城市中享有历史盛名的一些大门，当地的建筑形式和建筑材料也给设计者提供了参考。

建筑立面采用笔直的线性造型，用光感十足的白色石膏覆盖，这种石膏是当地建筑常使用的一种材料。入馆处的室外楼梯十分抢眼，它与美术馆在对角线处相交，与极简主义的建筑立面形成了令人赏心悦目的对照效果。

楼梯连通了街道和建筑的上层部分，同时将外立面切开，在建筑的内部结构和城市街道之间营造出视觉上的开放感。楼梯通往一个阶梯形的露台，可以作为公众的聚集地，也可作为表演和讲座的场所。透过长达5m的悬挑全玻璃幕墙，人们可以从画廊看到外面的公共露台。

里伯斯金工作室十分关注文化场所中公共空间的设计，这是他们设计中一直在强调的部分。MO美术馆虽位于密集的城市环境中，但几乎有四分之一的土地用于绿化设计。街道上正在新建一个雕塑花园，由立陶宛国家文化艺术奖的获奖者所创作的艺术装置作品——立陶宛别墅将在此展出。

MO美术馆的创始人兼收藏家维克托拉斯·布特库斯表示："里伯斯金工作室的设计极具象征意义和民主精神。本馆将成为立陶宛最大的私人美术馆，因此对我来说，美术馆向公众开放是很重要的，这样才能展现馆内藏品和本馆的精神内核。宽敞的公共空间非常好地展现了这种开放性。"他还说："这些公共空间贯穿了整个设计，其面积十分可观，从根本上为人们交流思想提供了方便。

在建筑的北侧，游客可以通过两层高的玻璃入口进入一个光线充盈的大厅。无论是在建筑的内部还是外部，建筑师都采用了几何形天窗这一共通的设计元素。一方面，光线能通过这些天窗进入底层空间，另一方面也使得上下层空间在视觉上形成连通效果，让空间整体性更好。

建筑师为藏品储存室设计了一面玻璃窗，这样就允许游客能够窥见里面还未展出的藏品。这些细节设计的奇思妙想让整个艺术馆饶有趣味。

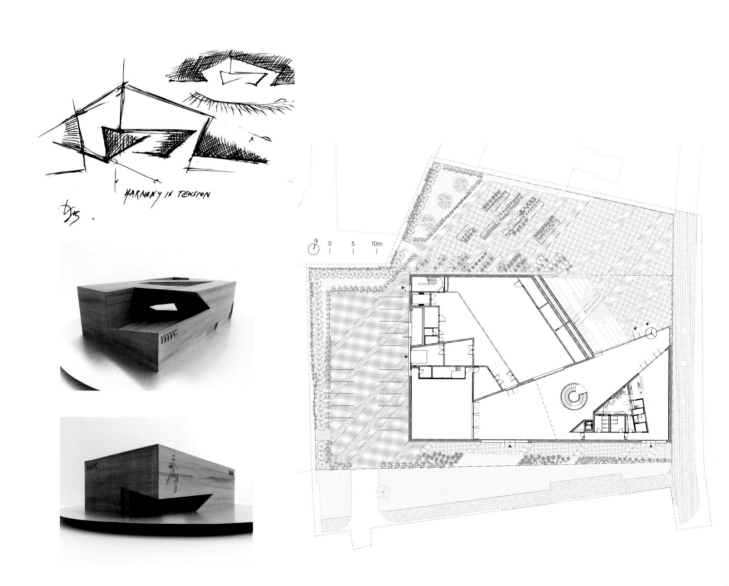

在入口附近，一条黑色的螺旋楼梯连接着主展厅和楼下的门厅，从而突出了博物馆的核心。展厅占地共1300m²，场地布局开放，有充分的空间举办永久和临时展览。

此外，博物馆还设有咖啡馆、书店、学习区、礼堂以及存储和行政空间。

"The MO Modern Art Museum is a cultural milestone for the city of Vilnius and Lithuania as a whole. This new world-class institution will showcase local art and will explore its links with the global art scene," said Museum Director Milda Ivanauskiene.

With a new public piazza located steps away from the historic medieval city, the 3,100m² museum stands as an expression of Vilnius' past and present. The MO Museum is conceived as a cultural "gateway" connecting the 18th century grid to the medieval walled city. The concept is inspired by the historic gates of the city and references the local architecture both in form and materials.

The rectilinear exterior facade is clad in luminous white plaster that references the local materials of the city. As visitors approach the museum, they encounter a dramatic open stair that intersects the museum on a diagonal axis, creating an expres-

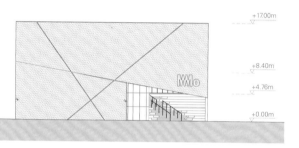

东立面 east elevation

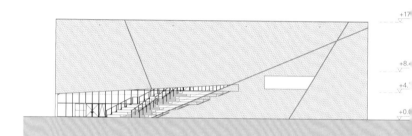

北立面 north elevation

sive counterpoint to the minimalist facade.

The stair cuts the facade open, connecting the street with the upper levels of the museum, and giving way to an openness that flows between inside and out. At the top of the exterior stair is a stepped open-air terraced roof that serves as a gathering area and place for public performances and talks. A five-meter, cantilevered, fully glazed wall allows views from the galleries to the public terrace.

An emphasis on including public space into their design for cultural institutions is one of the studio's most important defining goals. At the MO Museum, situated in a dense urban context, almost a quarter of the site is dedicated to green space. At street level, the installation Villa Lituania, created by the winners of the Lithuanian National Prize for Culture and Arts, was presented in a sculptural garden.

"One of the reasons I am drawn to Studio Libeskind's work is that it is both iconic and democratic," said Museum Founder and Collector, Viktoras Butkus. "The MO Museum is the largest private museum in Lithuania, so it is important to me that museum expresses openness and reflects the ethos of the

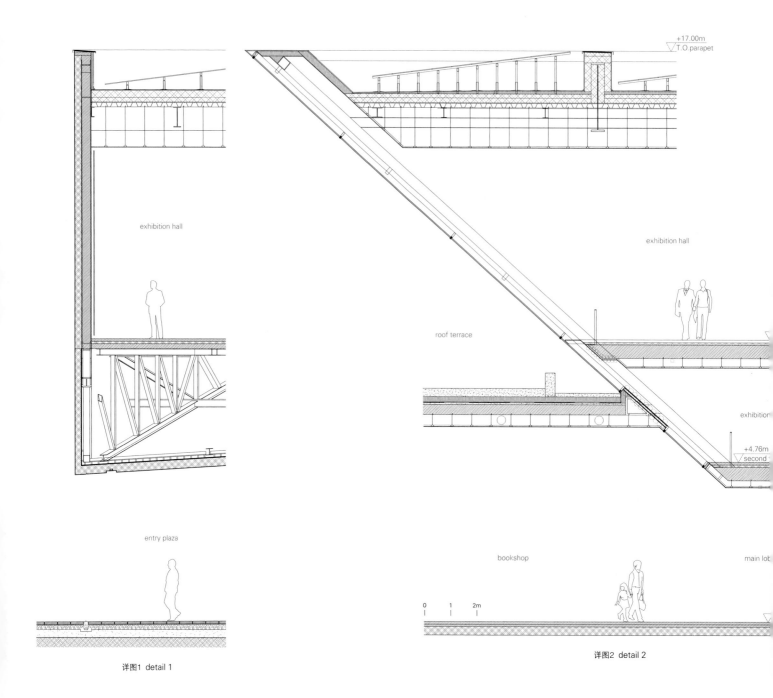

详图1 detail 1

详图2 detail 2

collection as well as the institution. The generous public spaces throughout the design play a vital role in communicating these ideas," added Butkus.

On the northern side of the building, visitors will enter though a two-story (8m-tall) glazed entrance into a light-filled lobby. Here, the theme of inside and outside continues to play out with geometric interior skylights that cut through the building, ushering in daylight to the lower floors and allowing views to the upper floors.

An interior glazed opening offers visitors a peek behind the scenes into the collection storage vault. These unexpected and sometimes playful moments continue throughout the museum. Near the entrance, a black spiral staircase connects the main gallery with the lower lobby and punctuates the museum's core. The galleries are laid out as open floor plans that provide 1,300m^2 of exhibition space dedicated to both permanent and temporary exhibitions of the museum's ongoing collection. The museum additionally includes a café, bookstore, educational areas and auditorium, as well as storage and administrative space.

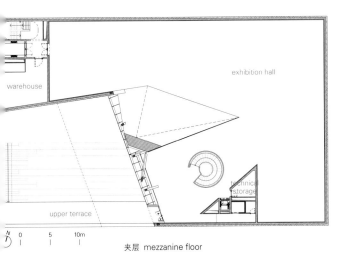
夹层 mezzanine floor

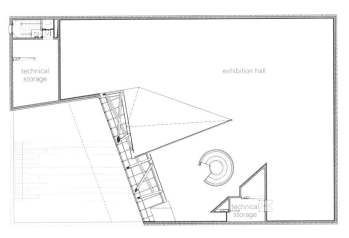
三层 third floor

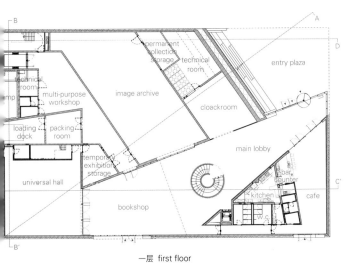
一层 first floor

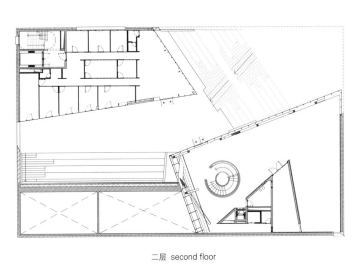
二层 second floor

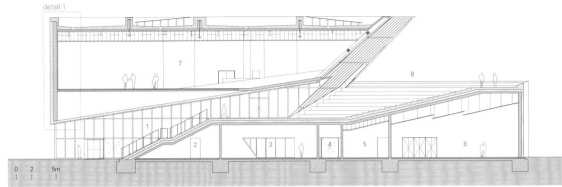

1. 入口广场 2. 衣帽间 3. 图像档案 4. 多功能工作坊 5. 临时展览储藏室
6. 大厅 7. 展厅 8. 上层露台
1. entry plaza 2. cloakroom 3. image archive 4. multi-purpose workshop
5. temporary exhibition storage 6. universal hall 7. exhibition hall 8. upper terrace
A-A' 剖面图 section A-A'

1. 热泵 2. 装货码头 3. 大厅 4. 设备储藏室 5. 库房
1. heat pump 2. loading dock 3. universal hall 4. technical storage 5. warehouse
B-B' 剖面图 section B-B'

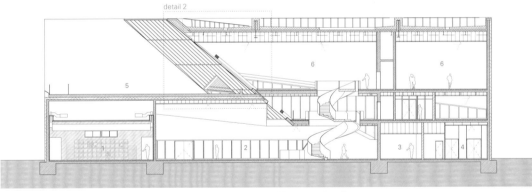

1. 大厅 2. 书店 3. 厨房 4. 咖啡店 5. 上层露台 6. 展厅
1. universal hall 2. bookshop 3. kitchen 4. cafe 5. upper terrace 6. exhibition hall
C-C' 剖面图 section C-C'

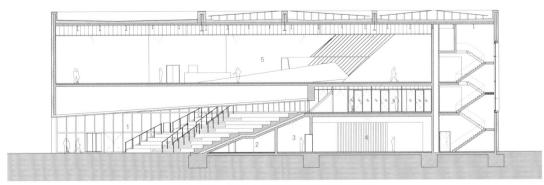

1. 入口广场 2. 设备间 3. 永久藏品储藏室 4. 图像档案 5. 展厅
1. entry plaza 2. technical room 3. permanent collection storage 4. image archive 5. exhibition hall
D-D' 剖面图 section D-D'

项目名称：MO Modern Art Museum
地点：Vilnius, Lithuania
设计建筑师：Studio Libeskind / 记录建筑师：Do Architects
总承包商：Naresta / 规划协调：Baltic Engineers
结构工程师：Ribinis büvis / 声学工程师：Akukon
电气工程师&装置：Baltic Engineers, APS
机械工程师&装置：Baltic Engineers, Prosfera, Folisita, Baltic System
立面工程师&装置：curtain walls – Glasma Service
客户：UAB MMC Projektai
总建筑面积：3,100m² / 高度：17m
材料：primary structure – cast-in-place concrete, prefabricated metal constructions; facade – glass/aluminum curtain wall, stucco; terrace – granite pavers; floors – polished concrete, wood (administration); interior walls – painted gypsum board; ceiling – perforated gypsum (ground floor hall, second floor exposition), metal mesh segmented ceiling (main exposition hall); spiral stair – prefabricated concrete stair steps, painted metal sheet railing, oak wood handle; windows – aluminum curtain wall, black mica matte paint finish; railings – stainless steel
造价：EUR 6.8 million / 设计时间：2015 / 施工时间：2017—2018
摄影师：©Hufton + Crow (courtesy of the architect) – p.151, p.153, p.155, p.156, p.157, p.158~159, p.162~163left; ©Norbert Tukaj (courtesy of the architect) – p.148~149, p.152, p.160, p.162~163right

都市里的村庄：城市的空间与规模

Village w:
Civic Spaces, C

据联合国报告显示，到2005年，全球50%的城市已经实现了城市化。如今，世界各地的大批移民都汇聚到城市，希望凭借城市生活的种种优势改善自己的未来，因此城市的转型速度正在不断加快。

城市密度、可持续性规划、城市活力，这些因素如何影响着公共空间与私人空间的关系？这是当前有关城市发展的争论点，围绕这个议题产生了下列问题：如何通过良好的设计打造出一个鼓

According to a United Nations' report, 50% of the world had become urbanized by 2005. Today, cities continue to transform at an ever-increasing rate, due to the migration of large, varied groups of people who hope to improve their futures through access to the benefits of urban life.

The present debate on cities focuses on the concerns about density, sustainable design and urban dynamics, and how such factors play into the relationships between public and private space. The debate has raised questions such as; how might we inhabit spaces which

深圳万科云城B2+A4区_B2 + A4 / Cloud City of Shenzhen Vanke / FCHA
深业上城LOFT_Shum Yip Upperhills Loft / Urbanus
绿洲排楼_Oasis Terraces / Serie Architects
海军部村庄_Kampung Admiralty / WOHA

都市里的村庄：城市的空间与规模_Village within a City: Civic Spaces, Civic Dimensions / Gihan Karunaratne

hin a City
ic Dimensions

舞人心的社区？我们对眼前一座城市的文脉应做何回应，才能给出一个宜时宜地、独树一帜的建筑方案？

本章举出了很多案例，它们针对新式的综合性建筑给出了解决方案，旨在用一种前卫的方式鼓励不同的社会群体见面与合作，为促成新的机遇、巩固社区间的联系带来了可能。

encourage collective confidence in the community through good design? How should we respond to an immediate urban context in order to ccreate architectural proposals that have a specific sense of place and time, as well as a unique identity?
There are various examples and architectural solutions in which new types of mixed-use, hybrid typologies, have been developed with the aim of encouraging distinct groups to meet and collaborate in novel ways that might create new opportunities and foster strong community bonds.

都市里的村庄：城市的空间与规模
Village within a City: Civic Spaces, Civic Dimensions

Gihan Karunaratne

  勒·柯布西耶对现代城市的景观影响深远，他彻底改变了我们的生活方式以及我们在建筑环境中的居住体验。勒·柯布西耶设计的马赛公寓是20世纪最具影响力的一座野兽派综合建筑，容纳了337套公寓、一座酒店、一个屋顶露台和两条购物街，堪称高密度住宅的范本。勒·柯布西耶将这个不可分割的社区称为"光辉城市"。马赛公寓重新构想了他"城中之城"的理念，以18层的混凝土结构重新定义了高密度住宅。这种建筑设计正应了勒·柯布西耶那句最有影响力的格言，"房子就是居住的机器"，于是这种风格的建筑规模逐渐扩大，应用到了整个社区。[1]

  柯布西耶认为，对于家园遭到第二次世界大战毁坏而流离失所的民众来说，建设高层公寓是解决他们安置问题的一个关键办法。此外，这种类形的建筑给城市带来了宽敞的住房空间，并将普通街道所包含的设施纳入到建筑当中。公寓由清水混凝土构成，在浇筑过程中使用木板为遮光板，给混凝土赋予了纹理。公寓室内狭窄，通常被布置成两层的复式单元，一端是客厅，占两层通高。楼层每户的纵深达21m (全楼宽24m)。为了节省通行空间，采用跃层连锁式布局，几对住户通过内部的小楼梯共用一条中央走廊。不像大多数公寓那样有一个"双叠"通道（即每个单元两侧各有一个通道），这栋公寓只需每三层设一条公共走廊，一共设5条，成为公寓里的"街道"。在建筑的室外设计方面，屋顶成为一个公共空间，供居民集会或锻炼。雕塑式的通风设备被一条跑道环绕，跑道中央是儿童泳池。"马赛公寓"被联合国教科文组织列为世界遗产，至今仍具有高度的影响力，是住宅类设计的典范。

  20世纪50年代是英国专注于战后国家重建和复兴的十年。那是一个对未来满载着理想、洋溢着希望的年代，一切将

So much of the modern cityscape is influenced by Le Corbusier. He revolutionized the way we live and the way we experience and inhabit the built environment. Le Corbusier's Unité d'Habitation in Marseille is one of the pre-eminent and most influential mixed-use Brutalist architectural schemes of the 20th century. Accommodating 337 apartments, a hotel, rooftop terrace and two shopping streets, the scheme is a template for high density housing – an integral community that Le Corbusier referred to as "the Radiant City". Unité d'Habitation redefined high-density housing by reimagining a "city within a city" housed inside an 18-story concrete structure. The architecture is Le Corbusier's most influential aphorism, "a house is a machine for living in", scaled up and applied to an entire community.[1]

Le Corbusier believed that the high-rise residential block was one of the key solutions for rehousing the masses displaced by the devastation of the Second World War. Furthermore, this building type could be used to create spacious urban housing with the same amenities as a typical street. Unité d'Habitation is constructed of exposed concrete given texture by the wooden boards used as shuttering formwork during the construction process. Inside, narrow apartments are typically positioned as two-story duplexes with a double-height living room at one end. One level of each apartment expands across the full 21-meter depth of the block, establishing a layout in which pairs of homes interlock around a central access corridor.

Unlike a typical apartment building, this arrangement meant that access corridors, known as "streets", were needed only on every third floor. There are five corridors in total. As regards the exterior, Le Corbusier designed the flat rooftop area as a common space in which occupants could assemble or exercise. Sculptural veiled ventilation stacks are encircled by a running track with a children's swimming pool in the center. Unité d'Habitation is listed by UNESCO as a world heritage site and remains one of the most highly influential residential design typologies.

帕克希尔庄园，地方议会修建的半废弃居民区，英国谢菲尔德
Park Hill, half-abandoned council housing estate, Sheffield, UK

在此重新开始。该时期的许多建筑师直接由地方议会聘用，设想出的住宅方案大胆激进。位于英国谢菲尔德的帕克希尔庄园就是其中之一，是公认的战后英国最为雄心勃勃的一个住宅项目。

遭到德国发动的闪电战之后，英国成千上万的人有的无家可归，有的蜗居在简陋的预制板房里。当时人民生活贫困，公共卫生条件差，公共设施不足，解决这些问题亟需政府增建新的救济住房、完善配套的基础设施。

中央政府多次提出要实现每年30多万廉租房的建设。无奈面临资金紧张、时间紧迫、物资匮乏的现实状况，政府需要找到一种新的建设方式。混凝土当选为主要的建筑材料。1951年的英国艺术节向公众展示了一个美丽新世界的蓝图。各地方政府试图将乌托邦式的现代主义思想转化为现实，对于伦敦郡和谢菲尔德市等最具开拓性的地方政府而言，公共住房将是其重建计划的重中之重。

勒·柯布西耶的马赛公寓当时吸引了很多建筑师的目光，一方面因为它那混凝土身躯彰显了现代主义的设计，另一方面它所包含的各种服务机构和生活设施满足了居民多方面的需求，给他们打造了一个舒适健康的居住环境。英国谢菲尔德市帕克希尔庄园的设计灵感就来自于马赛公寓。庄园里四座大型混凝土建筑矗立在山丘上，俯瞰这座约住着3000名居民的城市。它的设计师也效仿马赛公寓，把连通各个公寓的"街道"内置在建筑物中，这些"街道"蜿蜒向上，贯通到整个13层楼的屋顶。

新一代的建筑师、规划者和教育学家认为，建筑在解决社会问题方面可以发挥至关重要的作用，其核心目标就是建

The 1950s were a decade in which Britain focused on post-war national reconstruction and regeneration. It was a time of new beginnings – an age defined by a sense of idealism and hope for the future. There was a willingness to explore and implement new ways of doing things. Many architects of the period, employed directly by local councils, envisaged audacious and radical housing proposals. One such development was Park Hill Estate in Sheffield, widely considered to be one of the most ambitious residential developments in post-war Britain.

The Blitz had left thousands of people homeless or reduced to inhabiting poor and very basic prefabricated housing. There was a demand for new council houses with associated infrastructure to address the issues of poverty, substandard public health and inadequate facilities.

Numerous central government initiatives aimed to realize the construction of over 300,000 council homes per year. Money was tight, time was short, and materials were sparse: a new approach to construction was required. Concrete became the material of choice. The Festival of Britain in 1951 had presented the general public with a vision of a brave new world. Local governments would attempt to turn ideas of utopian modernism into reality. For the most pioneering councils, such as London County and Sheffield, public housing would be at the forefront of their redevelopment plans.

Le Corbusier's Unite d'Habitation held great appeal to many architects at the time, both in terms of its concrete construction and modernist design, and in its creation of a healthy environment, supported with many services and amenities that met residents' requirements. The scheme was an inspiration to Park Hill in Sheffield, where four large concrete buildings perched on a hill overlooking the city accommodated almost 3,000 residents. Park Hill's buildings consist of internal "streets" from which apartments are accessed, and which sweep upwards to 13 stories at the project's highest point.

巴比肯中心，英国伦敦
Barbican Center, London, UK

立社区。鉴于此，当代社区生活的新模式得到了充分详尽的考虑。"空中大街"是帕克希尔庄园最有名的一个概念性设计，它为居民提供了一个遮风挡雨的公共空间，不管遇到怎样的天气，居民都可以在这里闲庭信步，社交和娱乐。[2]

巴比肯住宅区位于伦敦，因在第二次世界大战中遭到德军的轰炸，破坏十分严重。战争结束，伦敦市政委员会决定采用先进的设计理念对这片区域进行开发和改造，雄心勃勃地想要以此次工程为范本，实现市中心生活模式的乌托邦理想。该项目方案想要具体地表现出遭到战争摧毁后的伦敦城浴火重生的新面貌，为此需要建设一座全国最高的塔式高层住宅楼，向世人展示伦敦如何进入到拥有英国最优秀的野兽派建筑典范的现代城市行列。

巴比肯住宅区是欧洲当时规模最大的基础设施项目，英国建筑事务所Chamberlin, Power & Bon负责该项目的设计。他们将2000多套公寓和附带的公共设施、市民文化区通过高架走廊连接在一起，实现了私人处所和公共空间的并置化布局。这个项目增强了建筑与地区历史的融合感，充分考虑了建筑与城市文脉的联系，将这些因素纳入到当时社会的首要任务中。这种紧扣主题的建筑模式在现代建筑师、规划师和地方政府看来也依然具有启发性。

巴比肯项目的设计方法以现代运动为基础，深受勒·柯布西耶的启发。建筑的诸多元素，譬如，彩色网格立面，从购物商场、人民广场到小型花园一应俱全的公共设施，共同营造了一种亲切的、个人化的体验。巴比肯项目的建筑立面由饰有纹理的现浇混凝土和悬挑阳台组成，上面大簇植物和叶子如瀑布般倾泻下来，把建筑衬得柔和了许多。该住宅区以城

A new generation of architects, planners and educationalists believed that architecture could play a vital role in dealing with society's problems. An overarching goal was to build communities.
Consequently, new models for contemporary community living were considered in great detail. The "street in the sky" was one of the most celebrated expression of the Park Hill design. It enabled residents to travel under cover, protected from the elements, and to meet socialize and play, in a communal setting.[2]
The Barbican housing estate is located in the city of London, an area that was extensively damaged by German bombing raids in WWII. At the end of the war the Corporation of London determined to redevelop and regenerate this part of the city by adopting progressive new design concepts. The ambition was to establish a showpiece development that would realize utopian ideals of inner-city living. The proposal would be a physical embodiment of London's rebirth from the destruction of the old city. The proposal was for a new city area incorporating the tallest residential tower in the nation; it illustrated how London could emerge as a modern city with one of the finest Brutalist architectural examples in the UK.
The Barbican was Europe's largest infrastructure project at the time. Designed by British firm Chamberlin, Powell and Bon, the development consists of over 2,000 residential apartments, together with public facilities and civic and cultural zones, all knitted together by elevated passageways. There is a juxtaposition of intimate and open spaces. A strong sense of historical incorporation within the site, and a considered relationship of scheme to context, are allied to a contemporary social agenda, a model of relevance which continued to offer inspiration to architects, planners and local governments until the present day.
The design approach pursued at the Barbican was grounded in the modern movement, predominately inspired by Le Corbusier. Architectural elements such as colorful gridded facades and public amenity, including open spaces ranging from plazas and squares to smaller gardens, give the estate an intimate,

深业上城LOFT，中国
Shum Yip Upperhills Loft, China

市中的职业人士为目标人群³，是最为成功的激进式公共发展项目之一，将居民住房和公共设施与设备聚合在一起。

"留仙洞"深圳万科云城（B2+A4区）（第172页）由坊城设计院（FCHA）设计，项目位于深圳南山区大沙河创新园区。该方案包含众多相互连接的分层结构和项目，包括办公室、共享公共空间和演讲厅。该项目既保证屋顶具有连续的公共绿地，又保证地下空间能提供良好的办公环境，表达了一种崭新的城市建筑方针，旨在解决中国现代城市的复杂性和不确定性。你要问设计方法？自然是"城中之城"的模式。

材料方面主要由清水混凝土、钢材和玻璃组成，适应了多样化的建筑比例和宽阔的建筑开口。该设计以高密度、多用途的都市化建筑形式，探索了社会、文化和商业的关系。B2+A4区属于混合类型的建筑地块，一系列经过分层的绿化带将生物多样性这个元素引入到项目开发中，以此鼓励各个年龄段的人们聚在一起交流、用餐和娱乐。

深业上城LOFT（第186页）是一个混合型建筑综合体，由六座摩天大楼组成，含公寓、办公楼和剧院。这个高端商业开发项目毗邻深圳CBD核心圈，美国SOM建筑事务所参与了部分项目设计。

综合体周围是一座小镇，毗连安静的步行街道。为了与这种构图呼应，项目设计单位都市实践建筑事务所创作了一个人工的"建筑之山"，该综合体包含了一些大型设施，在体量巨大的高层建筑和开口众多、阡陌交错的行人走廊之间进行了协调。整个系统把购物中心、办公楼和公寓组合在一起，建立了将居住、商业和文化空间融合到一起的新社区。各具特色

personal feel. The Barbican is dominated by elevations of in-situ cast textured concrete with ribbons of cantilevered balconies, softened by planting and foliage cascading down the façades. The Barbican, which was aimed at urban professionals[3], is one of the most successful radical public developments amalgamating housing and common public facilities and services under one roof.

B2 + A4 / Cloud City of Shenzhen Vanke (p.172) by FCHA is built for a creative design community in the Dashahe Innovation Zone in Nanshan District, Shenzhen. This scheme contains a multitude of connected layered structures and programs, including offices, shared public spaces and lecture halls. A balance is maintained between ensuring a continuity of the public landscape space on the roof and providing a suitable environment for underground office space. The project is an expression of a new urban architectural strategy intended to address the complexities and uncertainties in modern Chinese urbanism. The approach? Building a city inside a city.

In terms of materials, the design is composed mainly of exposed concrete, steel and glass, which responds to the varying building proportions and the wide scale of openings. The design which responds to rapid urbanization with a high density, mixed-use, urban typology, explores social, cultural and commercial relationships. Plot B2 + A4 is a hybrid architectural type which incorporates greenery in a series of layered levels to introduce an element of biodiversity into the development. It is a strategy that aims to encourage people, across all age groups, to come together to socialize, eat and play.

Shum Yip Upperhills Loft (p.186), a hybrid design complex featuring six skyscrapers comprising apartments, offices and a theater, is a fragment of a high-end commercial development adjacent to Shenzhen's central business district which was in part designed by SOM.

Urbanus has created "an artificial mountain" that responds to the structure of the neighboring composition by

绿洲排楼，新加坡
Oasis Terraces, Singapore

海军部村庄，新加坡
Kampung Admiralty, Singapore

的项目并置在一起后，一个混合型项目就此诞生。

该设计来自于一个令人兴奋而又激进的现代设计概念，呈现出了一系列舞台布景空间展现的效果。

项目设计大胆，功能齐全，共分为四个区域。A区是住宅区，由高耸的公寓楼群组成，每个公寓都有一个私人庭院用于社交活动。B区有一个酒店和办公空间，里面有一个封闭的花园和一个文化剧院。C区容纳了20栋的三四层高的写字楼，在基地中心形成一个村庄式的社区。D区，就是所谓的"建筑之山"，呈山形体量的建筑内部是一个展览中心。4

新加坡榜鹅社区的绿洲排楼（第202页）是由英国思锐建筑事务所设计的社区中心和综合医疗中心。排楼采用开放式框架结构，包括诊所、休闲区、零售店、餐馆、学习区和公共花园。位于中央的公共广场由一系列的花园排楼围成，用于鼓励居民在这里进行集体性的园艺活动。整个开发项目通过绿色的坡道和排楼连接起来，并附带零售店、教学中心、诊所等多种设施。

该建筑设计风格轻盈而开放；室内阳光充足，微风习习。南北和东西两向的主要通道十分宽敞，并附有天窗，与社区中心紧密相连。两条走道将空间自然划分成两片区域，一片是诊所等医疗相关区域，一片是商业区。5 这是一个很好的建筑范例，它促进了社会不同阶层和不同年龄段居民之间的社区凝聚力。

新加坡北部的海军部村庄（第216页）是面向老年人的公共住房开发项目，将医疗保健、社区空间、花园、公共设施和

creating a small town connected with quiet pavements. The complex's larger apparatus negotiates between the scale and volume of the high-rise and the intimate interconnections of porous pedestrian passageways. The scheme comprises shopping centers, offices and apartments, and establishes a new neighborhood which unifies habitation, business and culture. These distinctive programs are juxtaposed to encourage a mélange of cross programming.

The design is the result of an exciting and radical modern concept which provides for the unfolding of spaces in a series of theatrical set-pieces.

Designed in a bold, functional style, the scheme is split into four sectors. Zone A is residential and consists of loft apartments, each with a private courtyard for social activities. Zone B has a hotel and office with an enclosed inner garden and a theater complex for cultural events. Zone C accommodates 20 three- to four-story offices which form a small village community at the center of the site. Zone D, "the mountain", contains an exhibition center.[4]

Oasis Terraces (p.202), a Neighborhood Center and Polyclinic in Punggol, Singapore by Serie Architects, is an open frame structure comprising healthcare, leisure and retail facilities, together with restaurants, learning spaces and communal gardens. A central communal plaza framed by a series of garden terraces is designed to encourage community formation by functioning as a "horticultural project" for residents to maintain. The entire development is connected via a landscape of green ramps and terraces, enclosed by retail, educational and medical rooms and facilities.

The architecture is characterized by a sense of lightness and openness; daylight and breezes permeate the building. Wide, sky-lit north-south and east-west arteries closely connect the center to the surrounding neighborhood and naturally organize the space into a zone for the clinic and associated spaces, on the one

1. Amy Frearson, 'Brutalist buildings: Unité d'Habitation, Marseille by Le Corbusier', *Dezeen*, 15 September 2014, https://www.dezeen.com/2014/09/15/le-corbusier-unite-d-habitation-cite-radieuse-marseille-brutalist-architecture/
2. Britain's Post War Brutalist Architecture, The Park Hill Estate, presented by Tom Dyckhoff, https://www.youtube.com/channel/UCAaQgfEkcQQ0vxM0dIZLnFg
3. The Barbican: A Middle Class Council Estate, presented by Phineas Harper, Architecture Foundation, https://www.youtube.com/watch?v=FFDpqRxym_A
4. URBANUS encloses mixed-use Shenzhen development with 'artificial mountain volumes', *Designboom*, 26 June 2019, https://www.designboom.com/architecture/urbanus-shum-yip-upper-hills-loft-shenzhen-china-06-26-2019/
5. Oasis Terrace: Singapore's New Neighbourhood Centre and Polyclinic, *Archdaily*, 27 August 2015, https://www.archdaily.com/772537/oasis-terrace-singapores-new-neighborhood-center-and-polyclinic

商业设施垂直结合在一起。医疗保健和社区设施的提供巩固了社区的联系，响应了"积极老龄化"这一倡议，让老人以积极的态度颐养天年。

　　海军部村庄项目由新加坡国家公共住房管理机构——新加坡住房发展委员会发起，提供了100套老年人住房和300套年轻家庭住房，并配套了相关的服务设施。在较大型的综合社区中，如何给老年人安置住房成为当前普遍面临的困境。"Kampong"在马来语中意为"村庄"，该项目以此为核心理念，对解决该困境起到了模范推动作用。作为一个示范工程，如果成功，将对未来发展产生重要的影响。

　　以上所有案例均有一个统一的主题，即打造混合型建筑社区，突出设计的空间感、平衡感和开放感，以迎合广泛的住户群体和多种使用目的。每个项目在整体发展上都追求宁静的社区环境，采用分层的绿化方案，公共空间和私人空间并存。建筑师在设计上通过补充各种用途，演示了人们如何开启城中之城的生活模式，如何以积极互利的方式活跃在社区活动中。

　　这些项目从概念上讲都是打造在大城市中的"村落"。我们可以看到，当代建筑采用柔和的连续景观，对商业、居住和文化元素进行智能化分层设计，提供了一个有活力的居住环境。这种建筑设计，不论是对整个城市还是特定的社区，都做出了积极的贡献。

hand, and a commercial zone, on the other.⁵ This is a fine example of an architecture that fosters community cohesion between social classes and different generations, across multiple scales.

Kampung Admiralty (p.216), in the north of Singapore, is a public housing development for senior and elderly residents that combines healthcare, community spaces, gardens, public facilities and commercial amenities in a vertical configuration. The provision of healthcare and community facilities supports community bonding and promotes "active ageing".

Initiated by Singapore's Housing Development Board, the national public housing authority, Kampung Admiralty provides 100 units of elderly housing with 300 units for younger families, together with associated services. "Kampong" means "village" in Malay, and the village concept is at the heart of a scheme that sets a dynamic precedent of how to engage with the universal predicament of housing seniors within a larger, mixed community and is a prototype model that, if successful, could influence future developments.

All the above examples have a unifying theme, that is to say, they are hybrid architectural typologies aiming to accommodate a broad range of people and a wide range of uses, within designs characterized by a sense of space, balance and openness. Integral to each scheme are tranquil, layered landscaping schemes incorporated into the overall development. Public and private spaces exist side by side. Use complements the architecture in projects that demonstrate how people can live within an inner-city context and become actively engaged with the community in a positive and mutually beneficial manner.

Conceptually, these developments could be viewed as "villages" within the larger city. They show how contemporary architecture can provide a vital living environment through the softening of the continuous landscape and an intelligent layering of commercial, residential and cultural elements. This is an architecture which makes a positive contribution both to the wider city and particular communities.

深圳万科云城 B2+A4 区
B2 + A4 / Cloud City of Shenzhen Vanke

FCHA

本项目位于深圳南山区大沙河高新园区,也就是"留仙洞"总部所在的科技园北区的一块场地。万科想借此机会在北区绿色走廊地段打造一个集合万科上下游产业于一体的创意——设计园区。万科与项目设计单位——都市实践建筑事务所集中并筛选了各个设计单位的创意,最终确定在都市实践建筑事务所的指导下,通过集群设计的方式来完成第一个启动区域的设计。

坊城设计(FCHA)承担了A4+B2这两个地块的设计任务。这片区域呈刀形,长135m,宽50m,占地7500m²。场地南高北低,高差3.7m。依据城市设计方针,设计师用一个大面积绿化坡道将首层和负一层连接起来。整个地块被划分为连续的办公单元,北向区域拔地而起,成为一个独立的单元,以增加自然采光和通风口。既保证屋顶具有连续的公共绿地,又保证地下空间能提供良好的办公环境,在两者之间维持了平衡。

设计采用混凝土、钢材、玻璃等材料,使整个园区风格统一、空间丰富,而且产品多样化。

每个独立的办公单元都配有一个独立的、采光良好的观景台和户外休闲场地,供人独享清静、放松心情。

联合办公区位于B2区连接地面层和地下室的斜坡下面。室内总面积约2000m²,这里是整个项目的联合办公空间。各个产业和公司在这个巨大的联合办公空间内合作创新,实现了运营管理的高效和资源价值的共享。

内部的一个小型报告厅还兼顾报告、讲座、交流活动的功能。后方有一扇可以折叠的门,平常敞开,促进空间的流动,在举行讲座期间则保持关闭。

联合办公空间在支持相互交流的同时,也通过储物柜和书橱围出一些隔间,保证了空间的独立性。这种布置减少了上下层空间造成的视觉干扰,便于办公者在巨大的联合空间中顺利找到自己的位置。

独立办公空间位于光线充足的中庭,每一层可容纳30人,比联合区域更加安静、独立。

XFACTORY创客工坊配备了桌台宽大、工具齐全的工作间,有想法的创客们非常喜欢待在这里。设计师在具有错层高差的地方置入采光天井、交通楼梯及交流平台,增加了空间的变化性和趣味性。原始爆破留下的天然岩石墙面、采光良好的天井和通风走廊使这个区域更别具一格。

This project is located in the Dashahe Innovation Zone, in Nanshan District, Shenzhen; a site in the north district of the Science and Technology Park, for the headquarters of Liuxiandong. Property developers Vanke wanted to take this opportunity to begin a creative park/design community in the North Green Corridor that integrates upstream and downstream processes of the industry supply chain. Vanke and project planners Urbanus selected and convened the design units, and determined that the design of the first start-up area of the plot should be completed through Cluster Design under the guidance of Urbanus.

Architects FCHA undertook the design tasks of the two A4+B2 plots. The two plots cover an area of 7,500m²; the plots are knife-shaped, 135m by 50m in length and width. The site is high in the south and low in the north, giving a gentle slope with a height difference of 3.7m. Under the guidance of the urban design guidelines, the design connects the ground floor and the basement level through large green slopes. The plot was divided into continuous office units, with the northward

项目名称: B2+A4-Cloud City of Shenzhen Vanke / 地点: Nanshan District, ShenZhen, Guangdong, China / 建筑师: FCHA / 规划与景观设计: Urbanus / 智囊团: ICITY
主管合伙人: Zetao Chen / 项目主管: Xianzhi Zeng, Wan Gan / 设计团队: Jianhui Wang, XueFa Liu, Jie Liang, Jinlefu Su, JiaQi Zhao, Xusheng Guo, Xin Li
各组团设计师: A1+B3-URBANERGY; A2+B1-huayidesign; A3+B4-NODE / 施工图: CAPOL.Ltd. / 幕墙设计顾问: Shenzhen Tiansheng exterior wall consulting
客户: ShenZhen Vanke / 用地面积: 7,500m² / 设计开始时间: 2015.7 / 竣工时间: 2017.9 / 摄影师: ©Guanhong Chen (courtesy of the architect)

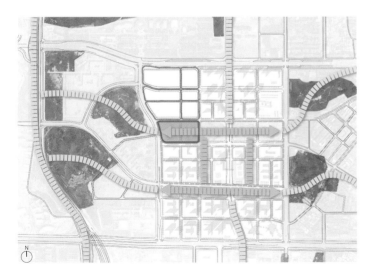

area raised as an independent building unit to increase natural lighting and ventilation openings. A balance is maintained between ensuring the continuity of the public green space on the roof and providing a good environment for below-ground office spaces.

The design makes use of concrete, steel, glass in order that the park should have a unified language, rich space and diverse products.

Each individual office unit has a separate, well-lit patio and outdoor leisure platform which provides a space for solitude and quiet relaxation.

The shared space is located underneath the inclined board connecting the ground floor and the basement level of one floor of the B2 block. The total interior building area is about 2,000m² and it is a co-working space for the entire project. Various industries and companies collaborate on innovation in this large shared space, operating efficiently, sharing resources, and sharing values.

A small lecture hall combines the functions of reports, lectures, and exchange activities. The folding doors that can be opened

at the rear are usually left open to encourage the flow of space, and can be closed during lectures.

The shared office platforms communicate with each other yet maintain their independence through the enclosures of cabinets and bookshelves. Users can find their own place within the large shared space with this reduced visual interference in the upper and lower spaces.

The independent office area is located in the well-lit atrium. Each floor of this section is a more quiet and independent office space. Each floor can accommodate about 30 people.

The large work desk and tool operation room provided by X-FACTORY maker workshop is a place where makers are happy to stay. Walls can be used for tool display and storage.

The design incorporates patios, traffic stairs and exchange platforms in places where there is a split-level height difference, creating a space that is both varied and interesting.

The natural rock wall left by the original blasting contributes to the uniqueness of this site, the light-well and the ventilated corridor also contribute to this uniqueness.

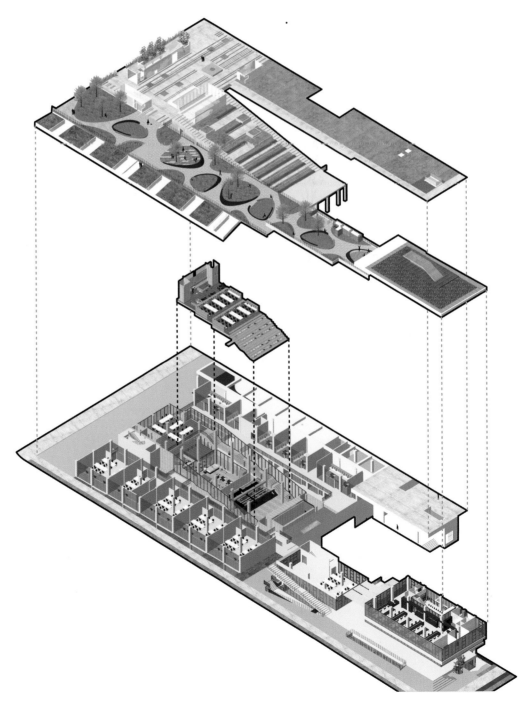

视频会议 video conference

小演讲厅 small lecture hall

开放办公区 open office area

采光天井 daylighting patio

自由办公区 free office area

小会议室 small meeting room
木材车间 wood workshop
金属车间 metal workshop
创客空间 maker space
办公单元 office units

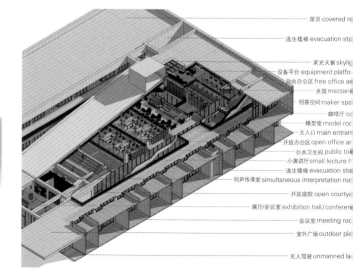

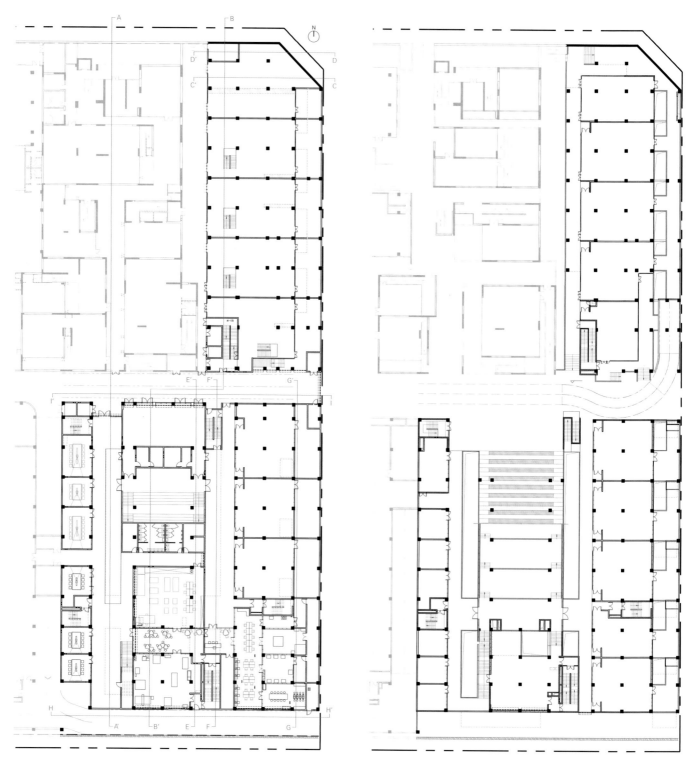

地下二层 second floor below ground   地下一层 first floor below ground

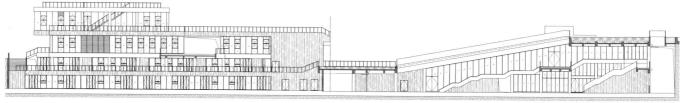

A-A' 剖面图 section A-A'

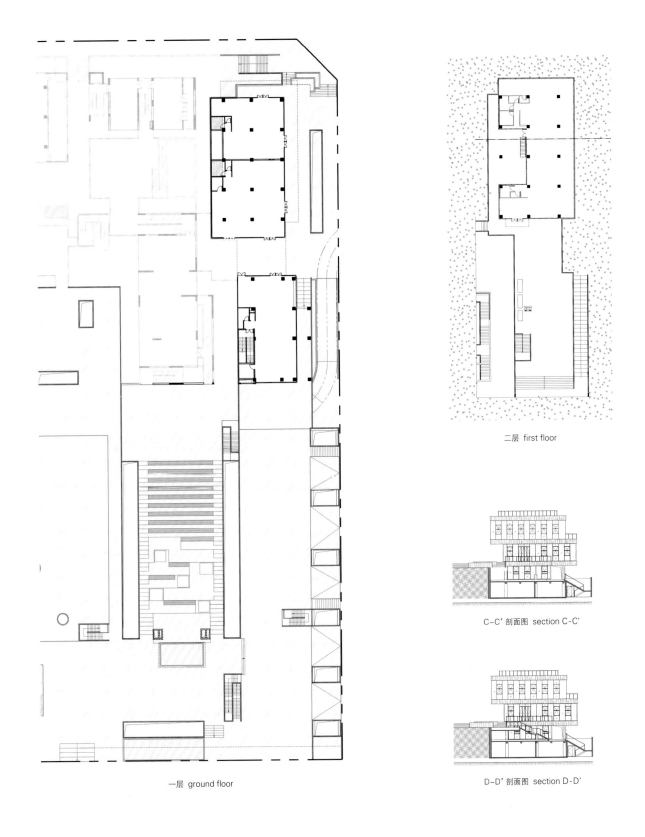

一层 ground floor

二层 first floor

C-C' 剖面图 section C-C'

D-D' 剖面图 section D-D'

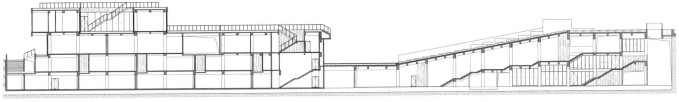

B-B' 剖面图 section B-B'

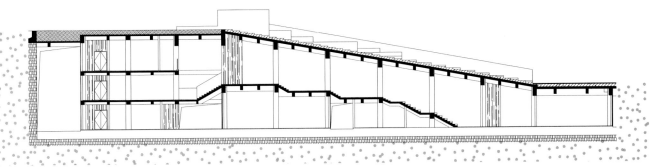

E-E' 剖面图 section E-E'

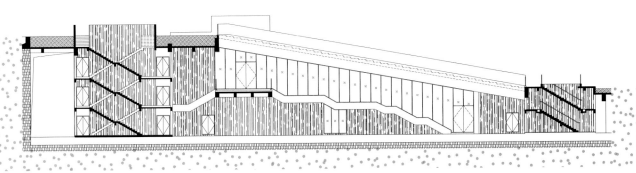

F-F' 剖面图 section F-F'

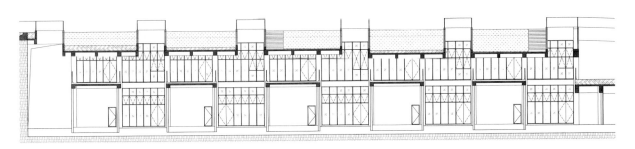

G-G' 剖面图 section G-G'

H-H' 剖面图 section H-H'

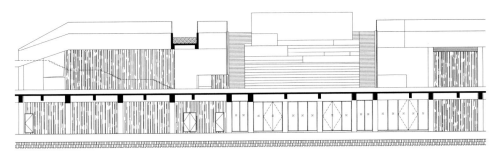

I-I' 剖面图 section I-I'

深业上城 LOFT
Shum Yip Upperhills Loft

Urbanus

深业上城LOFT这座高端商业综合体项目毗邻深圳CBD核心圈,位于两大市中心公园之间。它由6座高层建筑组成,内含办公、酒店、商务公寓等功能。这个项目由多家团队共同合作完成,都市实践建筑事务所作为其中的一个设计团队,接到的任务是在占地超过6万平方米的购物中心顶部,建造总面积为10万平方米的"阁楼(LOFT)",用于居住与办公。

为了缓解高层建筑垂直方向所产生的巨大压力,建筑师利用地块较大的平面面积,打造出了两座人工山形体量,将场地包围,并在自然形态上呼应了周边的莲花山和笔架山。同时,向内围合出一个安静的空间,以步行街道连接三四层的高密度办公"阁楼",排列出一个空间变化丰富的小镇,将阁楼剧场、展示交易中心等公共活动空间"装"入其中,逐渐将"庞大"而"坚实"的周围空间改造为"娇小"而"灵动"的内部区域。这座"阁楼之城"创造了一种新的聚落式生活模式,把居住、办公、商业与文化空间融为一体。

A区公寓"阁"的标准层高为9m,可灵活划分使用;于南北两侧设置外廊,具有良好的通风和防晒功能;户户拥有私家庭院,邻里之间既有交流,又保证了私密空间。立面呈灰白色"麻花"的格栅样式,力求将材料本身的性能发挥到极致。

B区酒店"阁"及办公"阁"两栋建筑向内围出一个庭院,庭院中央有一个黑色的盒子式小剧院,作为未来社区文化活动的发起点。立面采用灰白色陶板覆盖,其粗糙的质感带给人返朴归真的感受。

C区是办公"阁",由二十多栋3~4层的小型建筑聚合成小"村落",这些建筑呈组团分布,每个组团都拥有一个院子,每个组团的建筑之间的小巷宽4~6m,在它们之间还安插了一些露天广场,其中的街道宽度为8~15m。建筑组团在体量上穿插组合,产生了空中通道、露台、庭院、阳台、过廊等丰富的空间构筑物。

办公与商业平台之间以竖向交通连接,办公人群下楼即可漫步商业街,上楼又可继续工作,往北就是A区公寓"阁",往南就是D区展览中心。深业上城LOFT以新的建筑空间模式创造了多维度的生活方式。

D区位于场地南侧,是与A区相望的另一座建筑"山体"。这里原先是一座政府大楼,最初的功能被设定为办公总部。建筑体量的进深从26m到56m不等。建筑师在立面上设置了诸多开口,把庭院放进了建筑内部的各个区域,极大地丰富了室内外的空间层次。内庭、休息室以及室外的空中平台等各种空间在建筑内部都能连通,给整座建筑营造了多层次的空间氛围。

This high-end commercial complex is comprised of six high-rise towers containing offices, hotels and business apartments. Adjacent to the CBD district in Shenzhen, China, it is also located between two central parks. Urbanus' design task was to construct a 100,000m² "Loft", of apartments and offices, to sit on top of a shopping center larger than 60,000m².

To release the enormous pressure from the vertical dimension of the high-rise tower, the architects took advantage of the large area of the plot, creating two artificial mountain volumes, between which the site is enclosed. These connect the project to the natural forms of the surrounding Lianhua and Bijia Mountains. The design also generates quiet spaces; by connecting the 3-4 level high-density office Loft through sidewalks, it creates a small town with rich spatial variation. There are also some public spaces, such as the Loft Theater and the Trading and Exhibition Center that gradually trans-

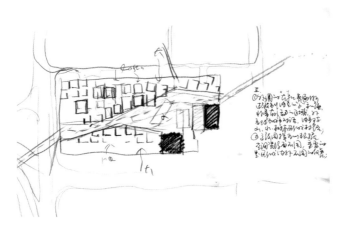
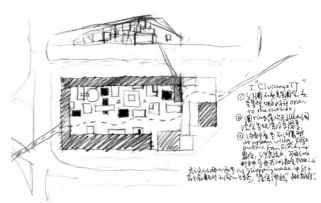

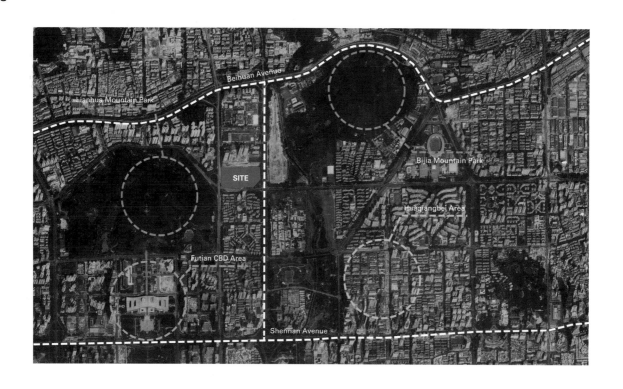

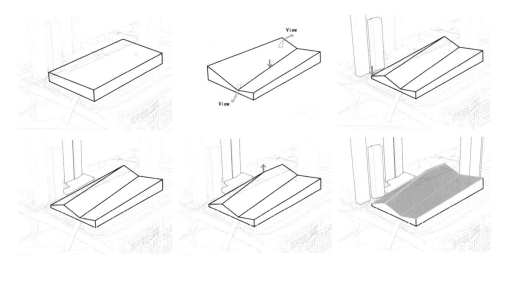

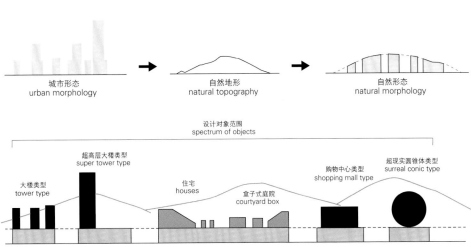

| 城市形态 | 自然地形 | 自然形态 |
| urban morphology | natural topography | natural morphology |

设计对象范围
spectrum of objects

| 大楼类型 | 超高层大楼类型 | 住宅 | 盒子式庭院 | 购物中心类型 | 超现实圆锥体类型 |
| tower type | super tower type | houses | courtyard box | shopping mall type | surreal conic type |

由裙楼和设计对象组成的城市
city of podiums and objects

项目名称：Shum Yip UpperHills Loft / 地点：Shenzhen, China / 建筑师：URBANUS / 主持建筑师：Meng Yan, Liu Xiaodu; Wang Hui – Interior / 项目总经理：Zhang Changwen, Lin Haibin, Jiang ling / 项目经理：Zhou Yalin, Zhang Xinfeng, Zhang Jiajia / 项目建筑师：Zhao Jia, Travis Bunt, Juliana Kei – Architecture; Zheng Na – Interior; Wei Zhijiao – Landscape / 团队：Su Yan, Zhang Haijun, Lin Junyi, Wang Yanping, Sun Yanhua, Zang Min, Cao Jian, Han Xiao, Zhang Ying, Wang Ping, Li Nian, Chen Guanhong, Yu Xinting, Xie Shengfen, Liu Kan, Silan Yip, Darren Kei, Sam Chan, Neo Wu, Danil Nagy, Daniel Fetcho, Yuan Nengchao, Lian Lili, Wang Lianpeng, Chen Hui, Zheng Zhi, Li Weibin, Milutin Cerovic – Architecture; Fang Xue, Liu Nini, Chen Biao, Li Xintong, Li Yongcai, Zhu Yuhao, Gao Jieyi, Chen Zhenzhen – Interior; Lin Ting, Zhang Yingyuan – Landscape; Xu Luoyi – Technical director; Wang Fang, Wang Yingzi, Wen Qianyue, Tang Disha, Guo Xusheng, Su Wushun, Tian Ye, Wang Jiahui, Tian Tao, Li Jiapei, Yu Kai, Shi Xianlin, Zhang Zhimin, Lin Xiaoyan – Internship / 结构、机械合作方：ARUP / 施工文件合作方：CAPOL / 室内设计合作方：Shenzhen Decoration and Construction Industrial co., Ltd. / 景观设计合作方：Shenzhen BLY Landscape & Architecture Planning & Design Institute / 立面设计合作方：Zhuhai Jingyi Glass Engineering / 照明设计顾问：Speirs + Major /Logo设计顾问：Corlette Design / 大楼、宴会厅建筑设计：SOM / 零售裙楼设计：ARQ / 客户：Shenzhen Kezhigu Investment Co., Ltd. / 用地面积：64,000m² (top area of shopping mall) 总建筑面积：105,000m² / 建筑楼层：3~14 / 建筑高度：36.45~67.5m / 设计时间：2012—2013 / 施工时间：2012—2018 / 摄影师：©ZtpVision (courtesy of the architect)

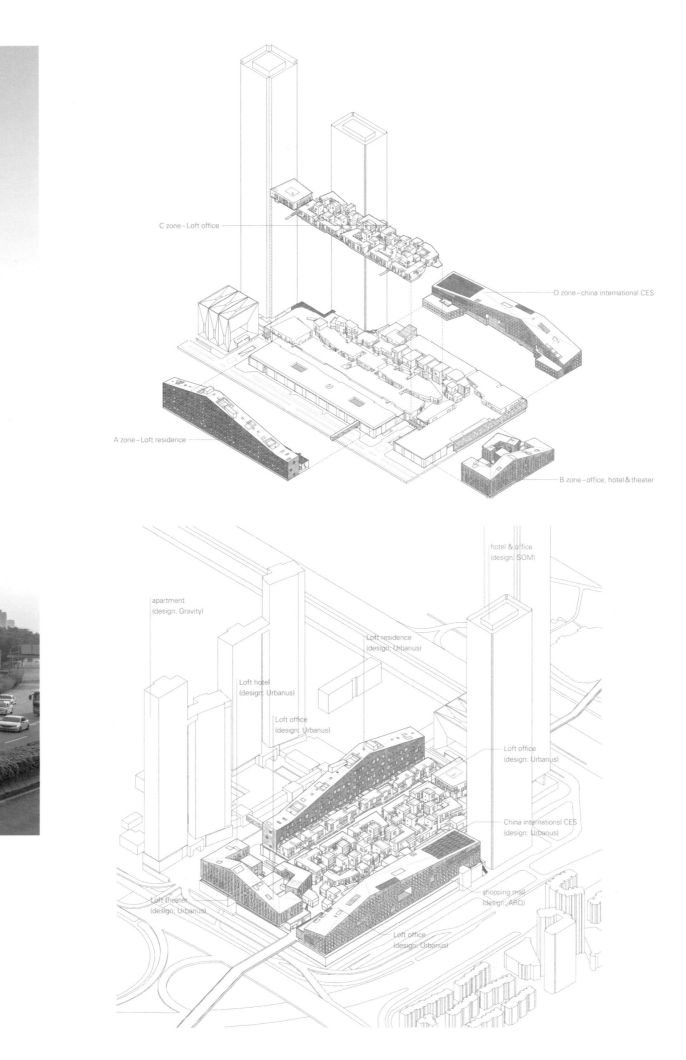

1. 住宅"阁"
2. 办公"阁"
3. 中国国际消费电子展示交易中心
4. 展览空间的挑空
5. 商业空间
6. 前厅
7. 剧场"阁"
8. 酒店"阁"

1. Loft residence
2. Loft office
3. CEEC
4. void above exhibition space
5. commercial
6. lobby
7. Loft theatre
8. Loft hotel

五层 fifth floor

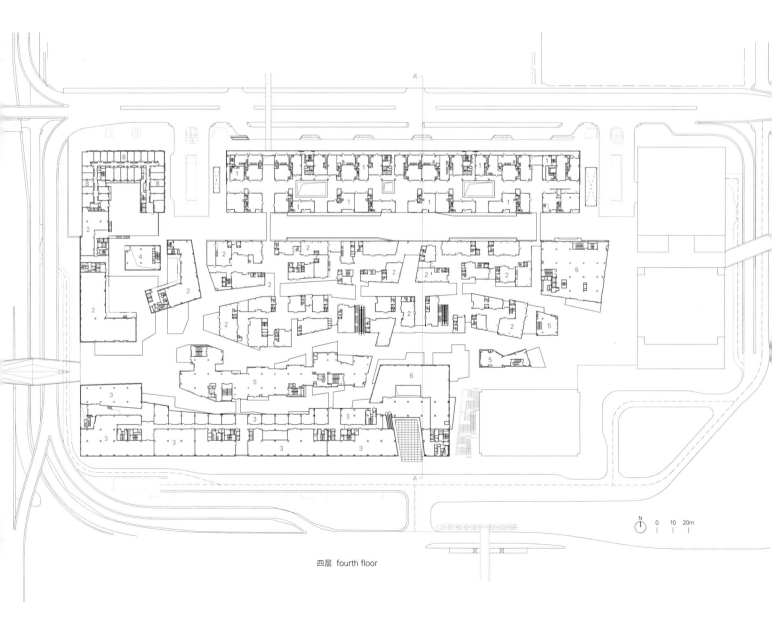

四层 fourth floor

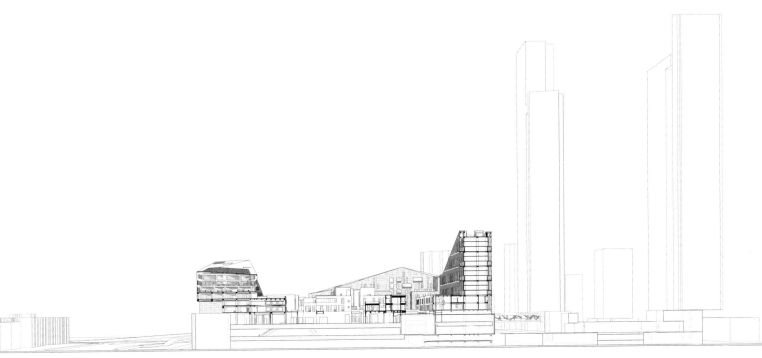

1. 前厅 2. 住宅"阁" 3. 办公"阁"　1. lobby　2. Loft residence　3. Loft office
A-A' 剖面图　section A-A'

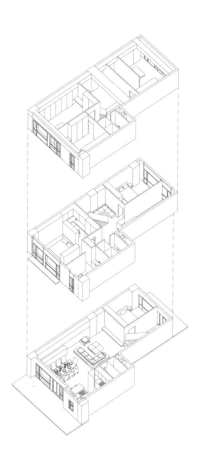 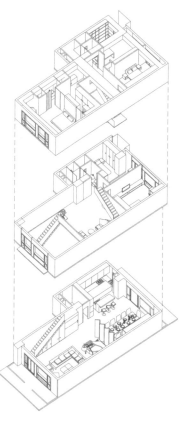 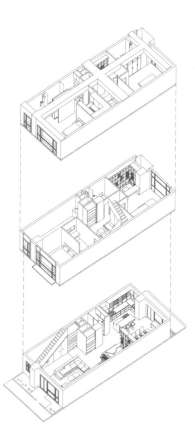

A区——住宅类型
zone A_residential types

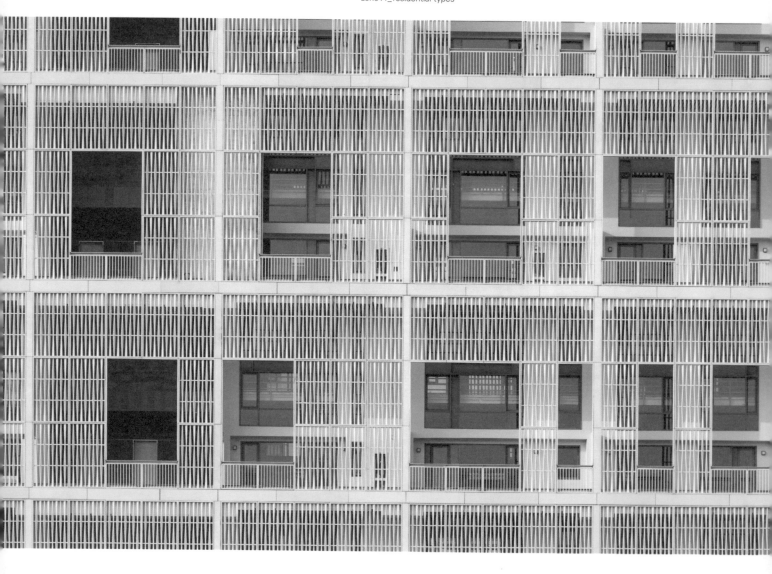

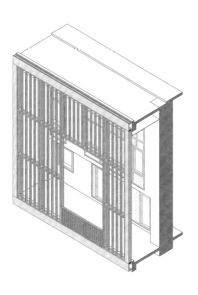

A区——南北立面单元
zone A _ south & north facade unit

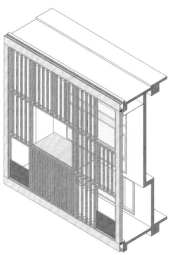

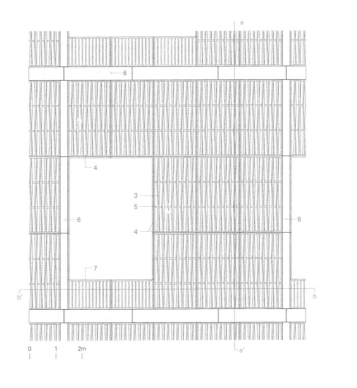

A区——南立面详图
zone A _ south facade detail

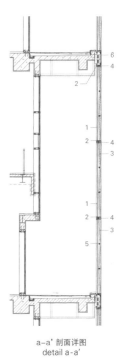

a-a' 剖面详图
detail a-a'

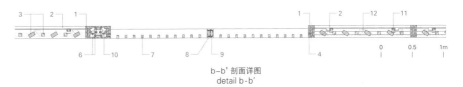

b-b' 剖面详图
detail b-b'

1. T-section steel column with epoxy and fluorocarbon coating 2. T-section steel beam with epoxy and fluorocarbon coating
3. UHPC grate 4. UHPC grate frame 5. UHPC grate cross-bar 6. 20mm thick UHPC panel with framing 7. UHPC railing
8. 80×16 galvanized steel column 9. 76×50×6 galvanized steel plate 10. 8# galvanized light channel steel 11. UHPC cap
12. 90×90×2 hard rubber sheet

form the "big" and "solid" periphery space to the "small" and "dynamic" inner region. This "Loft Town" creates a new model of settlement which integrates residents, offices, shopping malls and cultural spaces.

The standard floor height of the apartment loft in A District, is nine meters, which can be flexibly divided and used; verandas are set on the north and south sides, giving good ventilation and sun protection. Every household has a private courtyard, where neighbors can interact and socialize, while having their own private space. For the facade, we use a greyish twisted grid in order to maximize the performance of the material.

In B District, the hotel loft and office loft enclose an inner garden, where a black box theater is set to stimulate future cultural activities in this region. The facades are decorated with white ceramic plates with rough texture, giving people the feeling of getting back to nature.

In C District, there are over twenty office lofts of 3-4 stories, forming small "villages". These lofts were arranged by groups, each with a courtyard. The lanes inside have a width of 4-6m and between these groups are open spaces with streets of 8-15m width. Overhead channels, terraces, courtyards, balconies and galleries are interspersed among them, forming various spatial structures.

The office platform and business platform are connected by vertical transports. Workers can stroll in the mall downstairs, and conveniently return upstairs to work. In the north is a residential area, and in the south people can see exhibitions in D District. The Loft Town in UpperHills, by using a new building model, creates a multi-dimensional lifestyle for its residents.

D District is another "mountain" across from District A, located on the south side of the base. Originally a government building, it was initially set as the office headquarters. The depth of the building volume varies from 26m to 56m. By creating holes throughout the building's facades, the inner courtyards were set in different areas. This design greatly enriches the indoor and outdoor spaces. In the inner part of the building, aisles connect different levels of spaces, including inner courtyards, lounges and outdoor overhead platforms, creating a rich spatial experience.

绿洲排楼
Oasis Terraces

Serie Architects

新加坡榜鹅市的绿洲排楼是一个新建的社区中心+综合医院的建筑项目，由伦敦思锐建筑事务所与新加坡Multiply建筑师事务所联合设计而成。

该项目由新加坡住房和发展委员会（HDB）开发，旨在为其所在的社区提供服务，包括公共设施、购物商场、便利设施和政府综合医院。

思锐建筑事务所的设计引入了一个连续的花园平台，并使其呈阶梯状向水岸延伸。这些绿意盎然的花园成为重要的公共活动空间，包括各种儿童游乐场和一个天然的露天剧场。建筑师面临的最大挑战就是，采用一种既亲切而又令人惊奇的设计风格，将这些不同项目条理清晰地结合起来，让人一目了然。

20世纪70年代和80年代初，新加坡住房和发展委员会开发了一些在设计上颇具启发性的住宅楼，其中的公共走廊为开放式框架，通往各个公寓，在整栋建筑和街区形成了一个令人印象深刻的文明框架。绿洲排楼一层有各种空旷的露天平台，它们与翼缘墙相互连接，形成类似上述的公共门廊和柱廊，能够举办各种公共活动：婚礼、葬礼、麻将游戏，甚至是室内五人制足球赛。这些平台在设计上强烈地呼应了建筑师阿尔多·罗西1972年在米兰设计的加拉泰斯住宅项目。

建筑师将开放的结构体系转换成一个建筑框架——让建筑作为社交生活的容器。绿洲排楼的露天阳台由双柱支撑，环绕建筑的整个立面。它们一方面作为建筑重要的延伸空间，面向水岸，提供了绝佳的用餐地点。另一方面，它们作为综合医院的等候区，可供人在其上欣赏花园露台和水道景观。

建筑师用同样的双柱构造了一个16m高的带顶社交广场，并特意将其置于两条水道的交汇处，演绎着社区生活的一幕幕场景。广场与附近的绿洲轻轨站连接，成为社区中重要的交通节点。

建筑的每个立面都充斥着绿植，与环绕在餐厅和医院外部的阳台共同在室内和室外之间形成了一层"滤网"，使阳光和自然风柔和地透入内部空间，从而让整个建筑呈现出一种轻盈而开放的感觉。

屋顶同样覆盖以茂密的植被，包括用于城市农业的种植床。这些花园不仅能够为社区带来更加优美的景观，同时也能够促进居民从事集体性的园艺活动，通过种植和维护花园这类活动来建立更加牢固的社区纽带。

从更为抽象的层面上看，绿洲排楼展示了一种可能性，即对持久性体系结构的规则进行再次验证和转换。建筑师对已经在用途和时间上受到认可的建筑范式进行理论思考和概念重构，由此在不同时期的建筑作品之间建立了对话。他们将建筑创作转变为一种艺术实践，而这种艺术实践属于更广泛的工作和知识体系，重新确立了他们在建筑类型设计方面的卓越地位。

思锐建筑事务所的主创建筑师克里斯多夫·李表示："我们的设计借鉴了新加坡住房和发展委员会在20世纪七八十年代开发的住宅建筑中常见的开放式框架，并基于景观环境将这种框架转变为一个更加轻盈和开放的、能够捕捉和容纳各类社区功能的系统——这是一座为社区公共生活的展开而打造的建筑。"

Serie Architects, in collaboration with Multiply Architects, has completed Oasis Terraces, the new Punggol Neighborhood Center and Polyclinic in Singapore.

Oasis Terraces is a new generation of community center developed by Singapore's Housing and Development Board (HDB) to serve its public housing needs. It comprises communal facilities, shopping, amenities and a government polyclinic. Serie Architects' design utilizes a series of lush garden terraces that slope towards the waterway as one of the key elements to generate communal activities. These lush gardens act as communal spaces, children's playgrounds and a natural amphitheater. For the architects, the primary architectural challenge was to coherently bring these diverse programs together into a legible whole, with an architecture that is both familiar and surprising.

The architecture of HDB's housing blocks in the 1970s and early 1980s was instructive: the open frames of the communal corridors leading to individual apartments formed an impressive and civilizing framework for the housing blocks and the

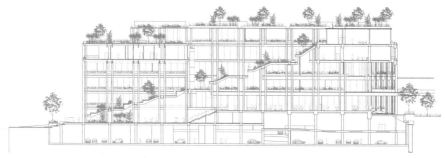

A-A' 剖面图 section A-A'

项目名称：Oasis Terraces / 地点：Punggol, Singapore / 设计建筑师：Serie Architects / 执行建筑师：Multiply Architects LLP / 土木结构工程：KTP Consultants Pte. Ltd. 机电工程：Bescon Consulting Engineers Pte. / 工程预算师：Northcroft Lim Consultants Pte. Ltd. / 消防工程：C2D Solutions Pte. Ltd. /立面：Aurecon Singapore Pte. Ltd. 景观：WNE Integrated Design Pte. Ltd. / ABC：Netatech Pte. Ltd. / 绿化标志：Afogreen Build Pte. Ltd. / 视听效果：Alpha Acoustics Engineering Pte. Ltd. / 照明：Light Cibles Pte. Ltd. / 室内设计：Multiply Interiors Pte. Ltd. / 卫生规划：DP Healthcare Pte. Ltd. / 项目管理：SIPM Consultants Pte. Ltd. / 总承包商：Rich Construction Company Pte. Ltd. / 打桩承包商：Keat Seng Piling Pte. Ltd. / 仪表承包商：Geo Application Engineers / 客户：Singapore Housing & Development Board / 用途：commercial, medical / 总建筑面积：27,000m² / 竣工时间：2018 / 摄影师：©Hufton + Crow (courtesy of the architect)

wider estates. The void decks on the ground floor, articulated and delineated by fin walls, resemble these communal porticos and colonnades, capable of hosting a diverse range of communal functions: weddings, funerals, mahjong games and even futsal. They strongly echo Aldo Rossi's 1972 Gallaratese Housing in Milan.

The architects transposed this grammar of the open frames into an architectural framework – a container of social life. Oasis Terraces' open verandas surround the entire elevation of the building, framed by double columns. These spaces provide the crucial spill-out areas for dining functions facing the waterfront and as breakout spaces for the polyclinic, which enjoys views of the garden terraces and waterways.

The same double columns also frame a 16m-tall covered social plaza, deliberately placed to front the junction of the two waterways. This plaza both contains and displays community events; all circulation through the site and from the adjacent Oasis LRT station culminates at the plaza.

Every visible elevation of the building is covered with lush planting. Together with the veranda spaces that wrap around the restaurants and polyclinic, the plants act as an environmental filter between the exterior and interior spaces. The architecture is also characterized by a sense of lightness and openness allowing daylight and breezes to permeate the building, promoting the use of natural ventilation.

The roof is also heavily landscaped and features planting beds for urban farming. The gardens play more than just an aesthetic role in the community; they are a collective horticultural project. By bringing residents together to plant, maintain and enjoy them, the gardens help nourish community bonds.

On a meta-critical level, Oasis Terraces demonstrates the possibility of revalidating and transposing the grammar of persistent architecture. Abstracting and repurposing architecture that has been sanctioned by use and time sets up a dialogue between the works of architecture of different periods, and turns the creation of architecture into an artistic practice that belongs to a wider body of work and knowledge, reasserting the primacy of working typologically.

Christopher Lee, Principal of Serie Architects said: "We've transformed the precedent of the 1970s' and 80's HDB blocks into a light and open frame that captures and accommodates diverse programs for the community in a landscape setting – it is an architectural framework for communal life to unfold".

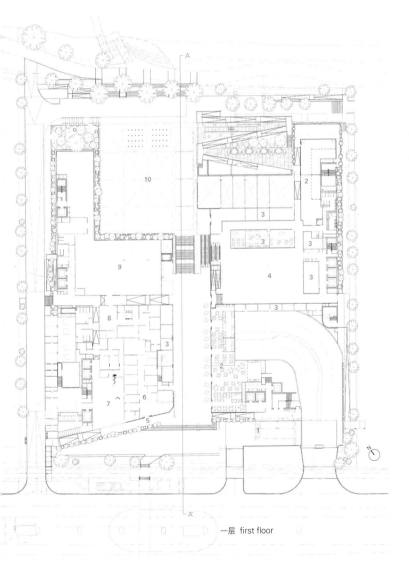

一层 first floor

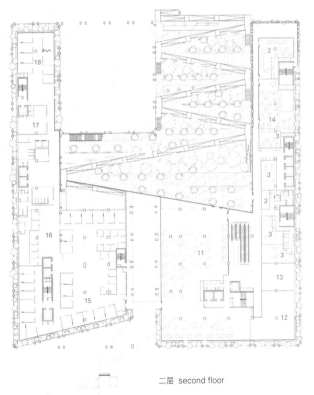

1. drop off/pick up 2. restaurant 3. shop 4. event space 5. entrance to polyclinic 6. rehab services 7. patient empowerment and display 8. X-ray 9. kidney dialysis center 10. community plaza below 11. food court 12. medical physiotherapy 13. pediatric clinic 14. retail 15. general chronic modules 16. lab 17. medical equipment store 18. enhanced AH & ECM

二层 second floor

1. 下车/上车处 2. 餐厅 3. 商店 4. 活动空间 5. 综合门诊入口 6. 康复服务 7. 患者授权和展示 8. X光室 9. 肾透析中心 10. 下方社区广场 11. 美食广场 12. 医学物理治疗 13. 儿科门诊 14. 零售 15. 普通慢性病诊室 16. 实验室 17. 医疗设备储藏室 18. 增强AH& ECM

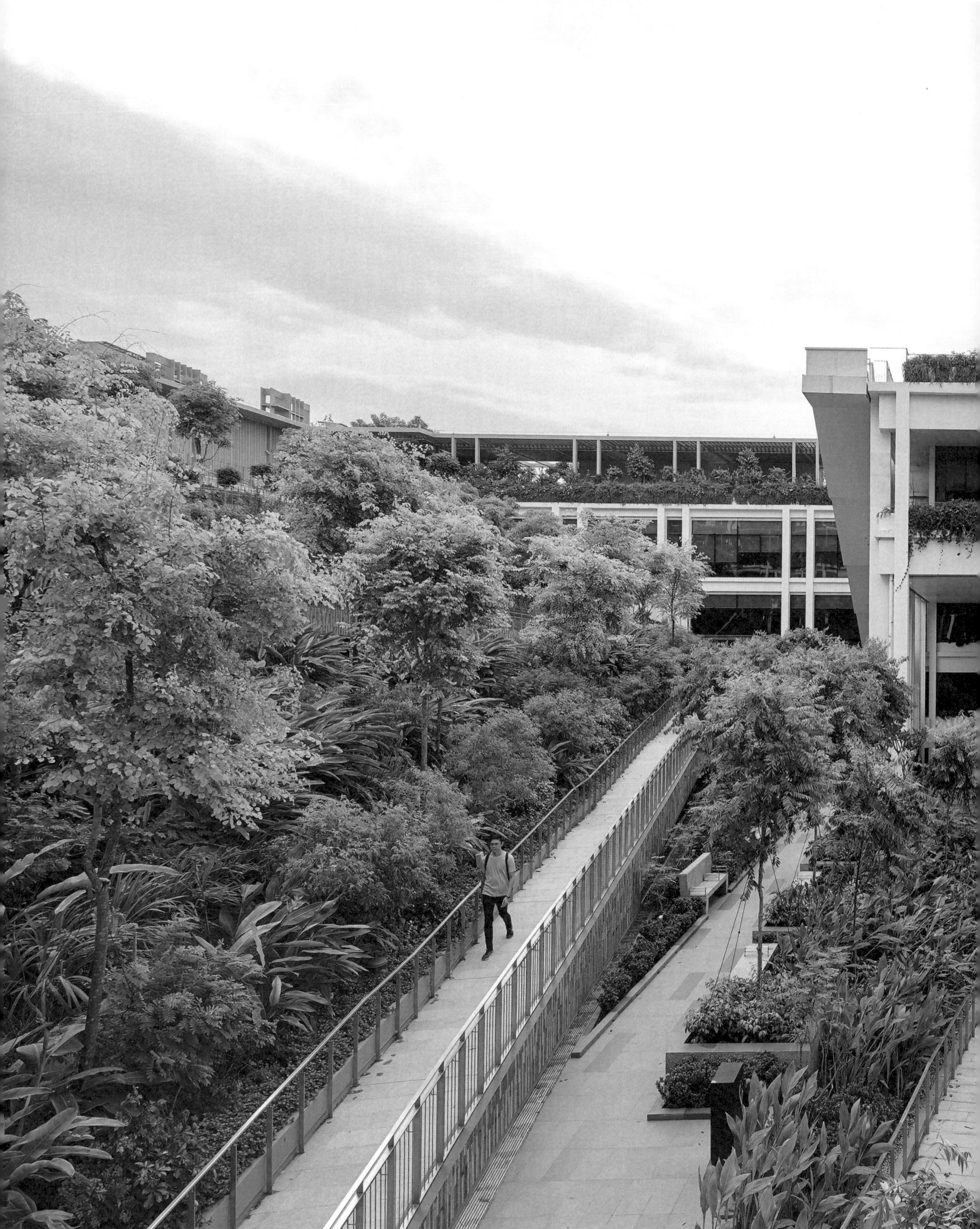

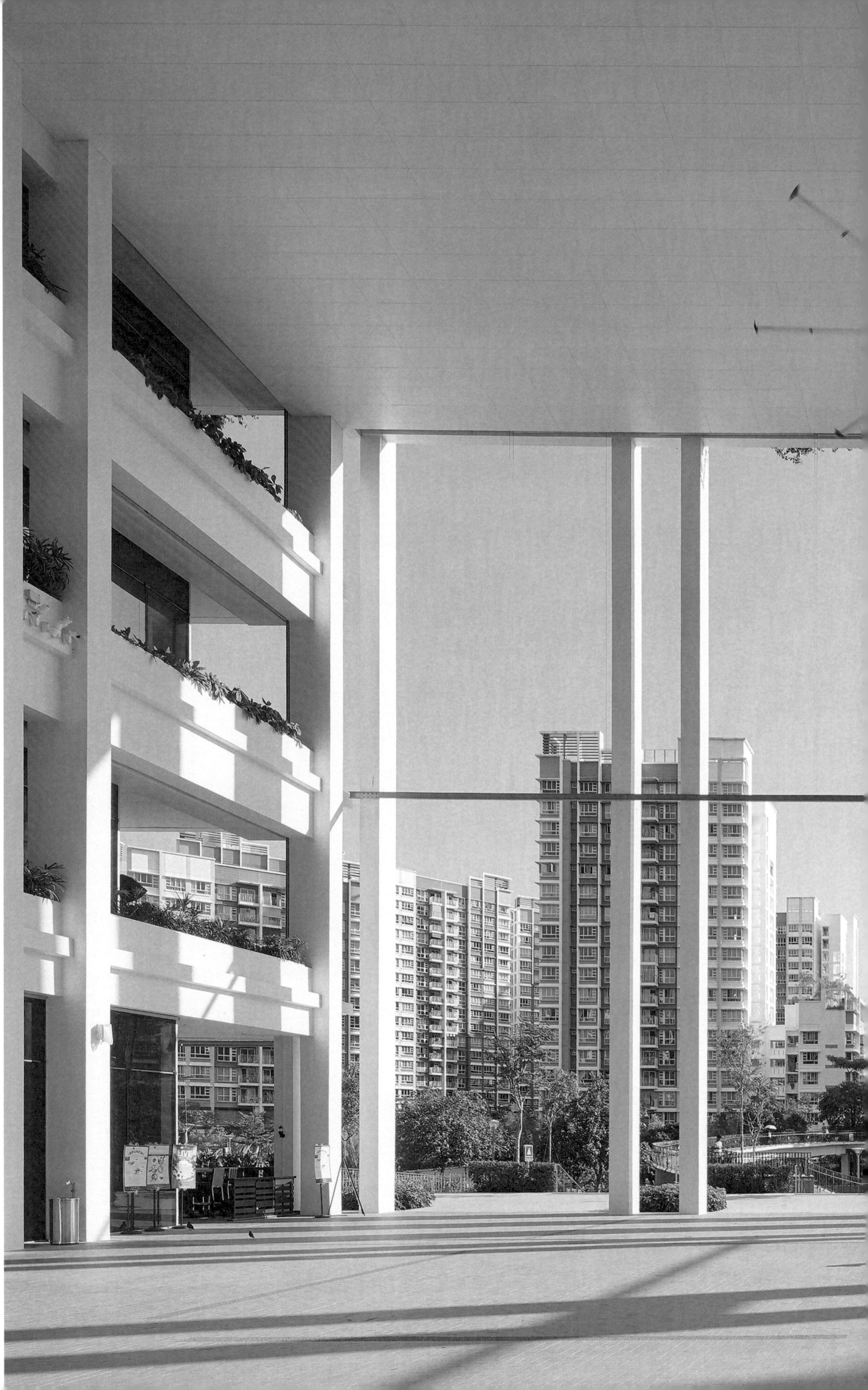

海军部村庄
Kampung Admiralty

WOHA

海军部村庄是新加坡首个将所有公共设施和服务空间融合在一个建筑体量里的公共建筑综合体。传统的项目做法是各个政府机构各自选取地块,建造几座各自独立、互不相连的建筑单体。该项目则不同,这个一站式的建筑综合体最大限度地利用了土地,为新加坡以老龄化人口为服务对象的建筑提供了范本。

项目位于一片只有0.9ha的基地上,用地面积十分紧张,此外还有45m的建筑高度限制,因此项目采用了一种类似于三明治的建筑形式。项目下层区域是一个社区广场,中层区域是一个医疗中心,上层区域则是社区公园和老年公寓,形成了一个垂直村落。这三层空间囊括了各种建筑功能,促进了功能空间之间的交叉性和多样性,同时将地面空间留给公众,方便他们从事各种活动。该设计拉近了医疗健康、社交活动、商业和其他便利设施之间的距离,加强了多代人之间的联系,支持"积极老龄化"这一倡议。

社区广场是这个社区综合体的"起居室",位于首层平面,是一个完全公共、具有极强渗透性的人行空间。在这个温馨、包容的空间内,公众可以参加一些有组织的活动和应季的节日庆典,可以尽情购物,也可以在二层的大排档餐饮区用餐。这个热带的广场通风极佳,上方有医疗中心为其遮风挡雨,无论雨天晴天,人们都可以自由活动。

在"村庄"中设置一个医疗中心意味着居民不再需要专门去医院看病,也不用因为时间和距离的原因被迫进行日间手术。

为了提升医疗健康服务水准,医疗中心的咨询和等候区采用自然采光,阳光透过四周的窗户和中庭照进室内,为患者提供了一个舒适的氛围。此外,位于上层区域的社区公园俯瞰着下层的社区广场,将老年人的生活与自然环境和各年龄层的人相连,消除了老年人的孤独感。

位于上层区域的社区公园是一个更为私密的垂直社区绿地,居民聚集在此,一起锻炼、聊天,甚至是共同打造一片社区农场。托儿所和

自带老年人护理服务的老年人活动中心等附属功能空间并排设置,将年轻人和老年人聚集在同一个空间内,为他们共同生活、吃饭和娱乐提供了条件。

这个综合体内部共设有104间老年公寓,分布在两个11层的建筑体量之中。共享入口处的"共享长凳"则鼓励老年人走出家门,与邻居互动。公寓内部采用了通用的设计原则,强调最大限度地实现自然通风和采光。

Kampung Admiralty is Singapore's first integrated public development that brings together a mix of public facilities and services under one roof. The traditional approach is for each government agency to carve out their own plot of land, resulting in several standalone buildings. This one-stop integrated complex, on the other hand, maximizes land use, and is a prototype for meeting the needs of Singapore's ageing population.

Located on a tight 9,000m² site with a height limit of 45m, the scheme is built using a layered "club sandwich" approach. A vertical "Kampung (village)" is devised, with a community plaza in the lower stratum, a medical center in the mid stratum, and a community park with apartments for seniors in the upper stratum. These three distinct stratums juxtapose the various building uses to foster diversity of cross-programming and frees up the ground level for activity generators. The close proximity to healthcare, social, commercial and other ameni-

项目名称：Kampung Admiralty / 地点：676 Woodlands Drive 71, Singapore 730676 / 建筑师：WOHA / 项目团队：Wong Mun Summ, Richard Hassell, Pearl Chee, Goh Soon Kim, Phua Hong Wei, Richard Kuppusamy, Jonathan Hooper, Yang Han, Lau Wannie, Gillian Hatch, Kwong Lay Lay, Zhou Yubai
室内设计：Sofwan, John Paul R Gonzales / 土木工程师：Ronnie & Koh Consultants Pte. Ltd. / 机电工程、绿化标志设计顾问：Aecom Pte. Ltd.
工程预算师：Davis Langdon KPK (Singapore) Pte. Ltd. / 景观设计顾问：Ramboll Studio Dreiseitl Singapore Pte. Ltd. / 总承包商：Lum Chang Building Contractors Pte Ltd. / 客户：Housing & Development Board / 用地面积：8,981m² / 建筑面积：53,066.49m² / 总建筑面积：32,331.60m² / 总楼面比率：3.6%
利用率：71.79% / 停车场：2 basement levels, with 252 car lots, 12 motorcycle lots and 364 automated bicycle parking lots / 设计启动时间：2013.3
打桩时间：2014.4—2014.10 / 施工时间：2014.10—2017.5 / 摄影师：©Darren Soh (courtesy of the architect) - p.220, p.221, p.222~223, p.228~229; ©K. Kopter (courtesy of the architect) - p.218, p.219; ©Patrick Bingham-Hall (courtesy of the architect) - p216~217, p.226

ties support inter-generational bonding and promote active aging in place.

The community plaza is a fully public, porous and pedestrianized ground plane, designed as a "community living room". Within this welcoming and inclusive space, the public can participate in organized events, join in the season's festivities, shop, or eat at the hawker center on the second story. The breezy tropical plaza is shaded and sheltered by the medical center above, allowing activities to continue regardless of rain or shine.

Locating a medical center in Kampung Admiralty means that residents need not go all the way to the hospital to consult a specialist for more routine day surgeries.

To promote wellness and healing, the center's consultation and waiting areas are bathed in natural daylight through perimeter windows and via a central courtyard. Views towards the community plaza below, and the community park above, also help senior residents feel connected to nature and to other people.

The community park is a more intimately scaled, elevated village green where residents can actively come together to exercise, chat or tend community farms. Complementary programs such as childcare and an Active Ageing Hub (including senior care) are located side by side, bringing young and old together to live, eat and play.

A total of 104 apartments are provided in two 11-story blocks for elderly singles or couples. "Buddy benches" at shared entrances encourage seniors to come out of their homes and interact with their neighbors. The units adopt universal design principles and are designed for natural cross-ventilation and optimum daylight.

1. 社区广场 2. 舞台 3. 生态池塘 4. 群落生境 5. 大道 6. 下车处 7. 电梯厅 8. 药房 9. 小吃店 10. 商店 11. 装卸货空间 12. 地下自行车停车处 13. 海军所在地 14. 集市 15. 操场 16. 熟食中心 17. M&E 18. 医疗中心 19. 庭院花园 20. 空中露台 21. 单间公寓 22. 多功能厅 23. 积极老龄化中心 24. 儿童保育中心 25. 健身区 26. 连桥 27. 社区农场

1. community plaza 2. stage 3. eco pond 4. biotope 5. thoroughfare 6. drop off 7. lift lobby 8. pharmacy 9. eating house 10. shop 11. loading unloading bay 12. underground bicycle storage 13. admiralty place 14. mart 15. playground 16. hawker center 17. M&E 18. medical center 19. courtyard garden 20. sky terrace 21. studio apartment 22. function hall 23. active aging hub 24. childcare center 25. fitness area 26. link bridge 27. community farm

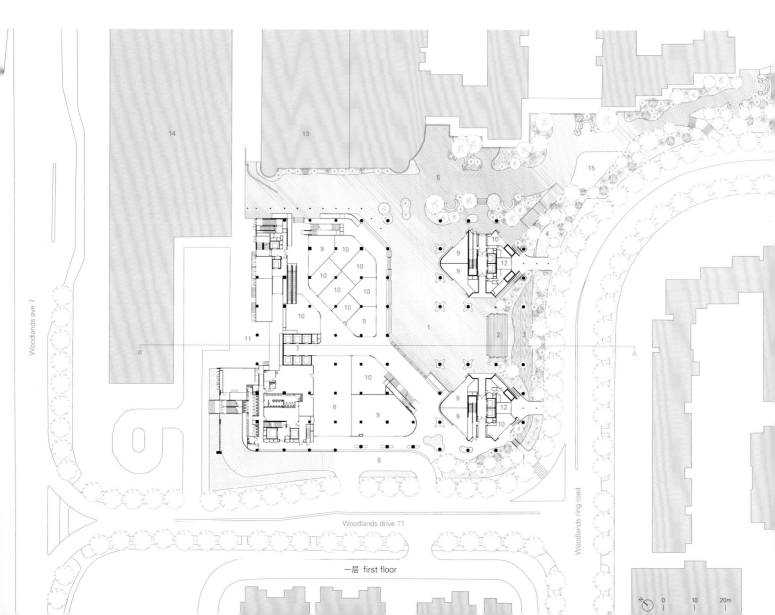

一层 first floor

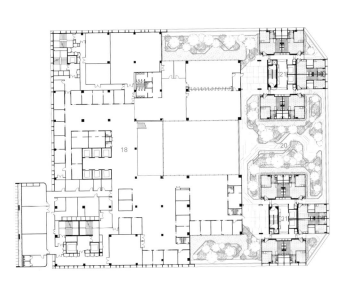

四层 fourth floor

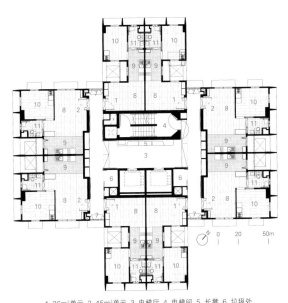

1. 36m² 单元 2. 45m² 单元 3. 电梯厅 4. 电梯间 5. 长凳 6. 垃圾处
7. 共享长凳 8. 起居室 9. 厨房 10. 卧室 11. 浴室

1. 36 sqm unit 2. 45 sqm unit 3. lift lobby 4. staircase shelter 5. bench 6. bin point
7. buddy bench 8. livingroom 9. kitchen 10. bedroom 11. bathroom

单元平面图 unit plan

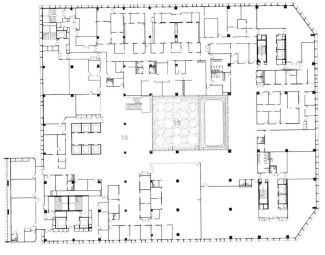

三层 third floor

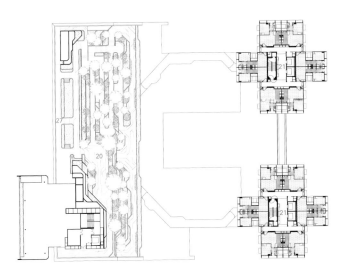

九层 ninth floor

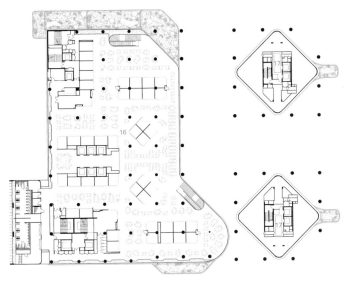

二层 second floor

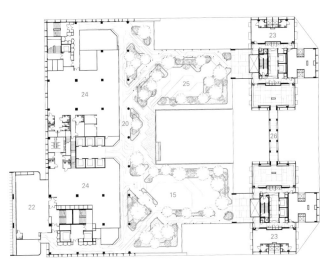

六层 sixth floor

南立面 south elevation

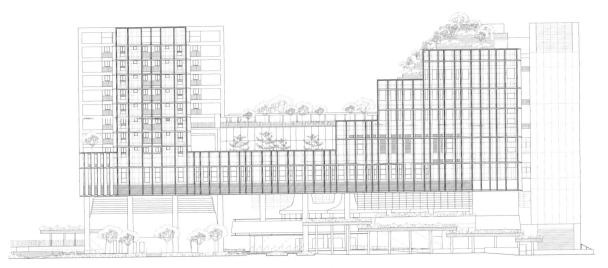

东立面 east elevation

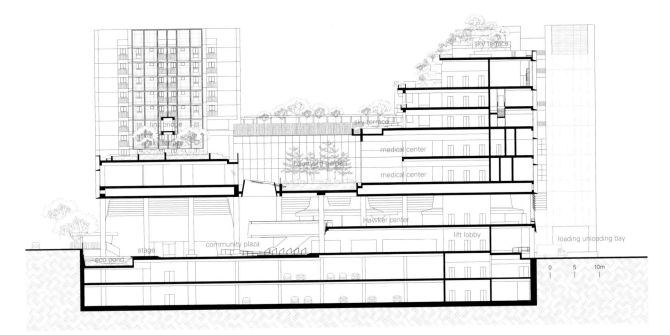

A-A' 剖面图 section A-A'

P8 Mannyoung Chung
Received his Ph.D. from the University of Seoul. Currently is Professor at the School of Architecture, Seoul National University of Science and Technology. His concerns range from the experiment in the design process to the theoretical intersection of architectural practice. He was the vice-president of the Korean Association of Architectural History.

P202 Serie Architects
Is an international practice specializing in architecture, urban design and research on the city. Founded in 2008 by Christopher Lee[right] and Kapil Gupta, Serie has offices in London, Singapore, and Mumbai. Martin Jameson[left] and Bolam Lee[center] are partners of Serie Architects.
Serie has gained a reputation for designing distinctive buildings in the public realm, with a special focus on cultural, civic and educational building. Has won a number of high-profile design competitions, including the BMW London Olympic Pavilion and the Singapore State Courts Complex. The practice's first Middle East project, the Jameel Arts Center in Dubai, was completed in 2018.

P34 Hyon-Sob Kim
Is Professor at Department of Architecture, Korea University. Studied modern architecture at the University of Sheffield, UK. Appointed as a professor at his alma mater Korea University in 2008, he has been teaching architectural history·theory·criticism and now focusing on writing a critical history of modern architecture in Korea. His recent publications include *Architecture Class: History of Western Modern Architecture* (2016), a Korean translation of *Building Ideas: An Introduction to Architectural Theory* (Jonathan Hale, 2017), "DDP Controversy and the Dilemma of H-Sang Seung's 'Landscript'" (2018) and "The Hanok Paradox: Modernity and Myth in the Revival of the Traditonal Korean House" (2019).

P186 Urbanus
Founded in 1999, under the leadership of partners, Xiaodu Liu[left], Yan Meng[center] and Hui Wang[right], it is recognized as one of the most influential architecture practices in China. Developed its branches in Beijing and Shenzhen. Received prominent awards, exhibited and published worldwide. Is now exploring opportunities for international and multidisciplinary collaborations to conduct a series of research projects focused on the contemporary urban China phenomena, including creative city development, post-urban village development, typologies for hyper-density and others. The projects have drawn international attention due to the firm's sensitivity to urban, historical, and social structures, its ability to integrate the potential resources in space and society, and its effectiveness in responding to the complexities of the urban environment.

P54 Kim Young-cheol
Graduated from the department of architecture in Korea University and the Graduate School, and studied the architectural theory of Schmarsow based on science of arts and German philosophy in the department of architectural theory, Technical University of Berlin. He is Professor at Ju-Si-Gyeong Liberal Arts College of Pai Chai University and acts as editor for the *Architectural Critics* in Korea and as moderator for the Saturday Architectural Theory Readers.

P74 Kang Howon
Born in Japan in 1964. Received a bachelor's degree from Hosei University, and master's degrees from Tokyo Metropolitan University and Columbia University. Had taught at Kyoto University of Art and Design, Tokyo Denki University, Kokushikan University, and Yonsei University. Since 2013, he has been an Associate Professor of Hongik University. Currently, he is a partner at Studio Void architecture+design. After five years of practice at Arata Isozaki & Associates, he has designed a variety of architectures based on Tokyo, Kobe, and Seoul as a method of 'rediscovering' architecture in a given environment and conditions, with his partner Kanehara Takaoki from 1999 to present.

P132 [eCV] estudio Claudio Vekstein_Opera Publica

Claudio Vekstein is an architect, founder of the international public architecture office [eCV]_Opera Publica, operating mainly in Argentina and Latin America since 1995, as well as in Arizona, USA. Is a full professor and researcher on Architecture and Public Infrastructure at Arizona State University since 2002. Recently, he has been Visiting Researcher at the Aalto University School of Arts, Design and Architecture in Espoo, Finland. His work received several awards including the Public Practices Award granted by the International Architecture Biennale of Argentina, the Baukunst Architecture Award granted by the Stiftung Städelschule für Baukunst, Germany, etc.

P148 Studio Libeskind

Studio Libeskind was established by Polish-American architect, Daniel and his partner Nina Libeskind in Berlin, Germany in 1989 after winning the competition to build the Jewish Museum Berlin. In 2003, Studio Libeskind moved its headquarters from Berlin to New York City to oversee the master planner for the World Trade Center redevelopment. The Studio has offices in New York and Zurich. Creates architecture that is resonant, practical and sustainable. The Studio has completed buildings that range from museums and concert halls to convention centers, university buildings, hotels, shopping centers and residential towers.

P172 FCHA

Is a research based architectural practice with a special interest in the pattern of Chinese urbanisation. Its team with international background investigates and explores Chinese contemporary architecture in mass urbanised locations through study and research of Chinese cities.
Founding Partner and Creative Director, Zetao Chen received his Master's Degree from the Berlage Institute. Is a Member of Shenzhen Architectural Design Review Expert Bank. Has been invited to participate in the research of design topics of UABB(Bi-City Biennale of Urbanism Architecture) for many times. Many works have won prizes in international competitions and have been awarded by many professions institution at home and abroad.

P86 EDDEA

Is a 20 years experienced Spanish Architecture firm that employs about 40 staffs from different nations. Led by a team of partners, currently works for a wide range of national and international projects. Its biggest interests are architecture and urban planning, which joined with the best strategy thinking. Works to create the conditions for a new dialog between Society, Nature and Economy, through a holistic development targeting strategy. Has built up the necessary experience to work in any country, having demonstrated absolute trust and the ability to reply to the idiosyncrasy of each country in the work they have carried out in many countries. Projects have been exhibited at the 11th International Architecture Exhibition at the Venice Biennale among other events. (Ignacio Laguillo, Luis Ybarra, Harald Schonegger, Jose Luis Lopez, Jose Maria de Cardenas, from the left, in the picture)

P116 Steimle Architekten

Thomas Steimle[right] (1974) and Christine Steimle[left] (1975) received their diploma of architecture at the University of Stuttgart in 2000. Both of them had worked at Wulf Architekten. Thomas founded Steimle Architekten in 2009. He has been teaching building construction and design at HFT Stuttgart since 2009. Became a member of the Association of German Architects (BDA) in 2013 and has been a member of the advisory board at the Council of Architecture in Baden-Wuerttemberg since 2017. Christine has been teaching design at the University of Stuttgart since 2007. She joined Steimle Architekten in 2014.

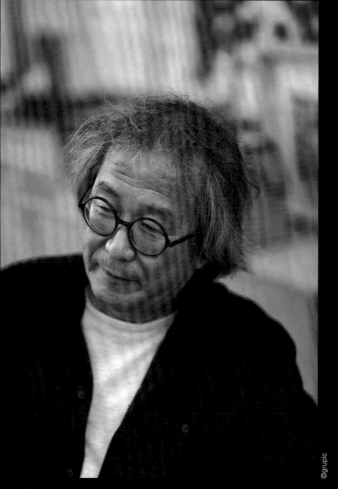

P4 Richard Ingersoll
Born in California, 1949, earned a doctorate in architectural history at UC Berkeley, and was a tenured associate professor at Rice University (Houston) from 1986-97. He has lived off and on in Tuscany since 1970 and currently teaches at Syracuse University in Florence (Italy), and the Politecnico in Milan. He was the executive editor of *Design Book Review* from 1983 to 1997. His recent publications include: *World Architecture, A Cross-Cultural History* (2013); *Sprawltown, Looking for the City on its Edge* (2006); *World Architecture, 1900-2000; A Critical Mosaic, Volume I: North America, USA and Canada* (2000). He frequently writes criticism for *Arquitectura Viva*, *Architect*, *Lotus* and *C3*.

P18 Seung, H-Sang
Born in 1952, studied at Seoul National University. Worked for Kim Swoo-geun from 1974 to 1989 and established his office 'IROJE architects & planners' in 1989. Published several books including *Beauty of Poverty* (1996, Mikunsa), and taught at North London University, Seoul National University and Korea National University of Art, and TU Wien as visiting professor. The America Institute of Architects invested him with Honorary Fellow of AIA in 2002, and Korea National Museum of Contemporary of Art selected him as 'The Artist of Year 2002'. In 2007, Korean government honored him with 'Korea Award for Art and Culture', and he was director for Gwangju Design Biennale 2011 after for Korean Pavilion of Venice Biennale 2008. Was the first City Architect of Seoul Metropolitan Government in 2014 and finished its term in 2016. And now he is Chair Professor at Dong-A University and working as Chief Commissioner of Presidential Commission on Architecture Policy.

P58 IDMM Architects
Heesoo Kwak graduated from Hongik University and established IDMM Architects in 2003, Seoul. Pays attention to the constantly occurring urban phenomena and extracts issues that need to be decoded and presents them through new types of architecture, exhibitions, cartoons, and contributions. Writes a column in Korea JoongAng Daily, titled "Architect Heesoo Kwak's sketch of city". Has been awarded the 2016 American Architecture Gold Prize, the 22/24/25th World Architecture Award, the 2017 A-Awards (ARENA HOMME+), the 2016 Hongik Honored Alumni Award, the 39th Korean Institute of Architects Award, Korean Architecture Award- Grand Prize in 2016, and Korean Architecture Award-Prime Minister's Award in 2008, 2012, and 2018.

P216 WOHA
Was founded by Wong Mun Summ and Richard Hassell in 1994 in Singapore. Mun Summ Wong graduated with Honours from the National University of Singapore in 1989. Is a Professor in Practice at his alma mater, at the Department of Architecture. Sits on the Nominating Committee of the Lee Kuan Yew World City Prize in Singapore. Richard Hassell graduated from the University of Western Australia in 1989, and was awarded a Master of Architecture degree from RMIT University, Melbourne, in 2002. Has served as an Adjunct Professor at the University of Technology Sydney, and the University of Western Australia.

P78 **Herbert Wright**
Graduated in Physics and Astrophysics from London University, and has worked in software publishing, press analysis and arts administration. Writes about architecture, urbanism, and art for magazines and newspapers across Europe. Is contributing editor of UK architecture magazine *Blueprint*. Launched and curated Lisbon's first public architectural event 'Open House' in 2012. His first book *London High* (2006) was about London high-rise, and later book projects include collaborations with Dutch architects Mecanoo and Expo 2015 Gold Prize designer Wolfgang Buttress. Delivers occasional talks, tours and discussions.

P164 **Gihan Karunaratne**
Is a British architect and a graduate of Royal College of Arts and Bartlett School of Architecture. Has taught and lectured in architecture and urban design in UK, Sri Lanka and China. Writes and researches extensively on art, architecture and urban design. Currently is the Director of architecture for Colombo Art Biennale 2016. Has exhibited in Colombo art Biennale in 2014, Rotterdam Architecture Biennale in 2009 and the Royal Academy Summer Exhibition. Is a recipient of The Bovis and Architect Journal award for architecture and was made a Fellow of Royal Society of Arts (RSA) for Architecture, Design and Education in 2012.

P104 **Studio Seilern Architects**
Is a London based international creative practice established in 2006 by Christina Seilern with the intent of producing exceptional architecture that lasts, working across geographies, building sizes and typologies. Diverse portfolio of built work spans the UK, Europe, North Africa and the Middle East. Believes that architecture does not just rely on innovation and design, but on the ability to deliver a project from conception through to completion. Its expertise lies within commercial, high-end residential, mixed-use, cultural, masterplanning and educational projects.
Christina is a regular juror for the AR MIPIM Awards, World Architecture Festival, the RIBA EyeLine, LEAF Awards and PAD London.

P38 **BCHO Partners**
Was opened by Byoungsoo Cho in 1994. He has been actively practicipating with themes such as 'experience and perception', 'existing and existed', 'I shaped house and L shaped house', 'contemporary vernacular'. Has taught the theory and design of architecture at several universities including Harvard University, Columbia University and University of Hawaii. Is the recipient of Architizer A+ Awards and AIA Northwest and Pacific Region Design Awards in 2015, KIA Award in 2014 with several previous AIA Honor Awards in Montana Chapter and in N.W. Pacific Regional. His projects has been published in different magazines including *The Architectural Reveiw*, *Dwell*, and *deutsche bauzeitung(db)*.
Ji-hyun Lee graduated in Industrial Design from KAIST, and Architecture from Politecnico di Milano, Italy. Ja-yoon Yoon has received her Bachelor of Architecture at Korea University and her Masters of Interior Design at Royal College of Art. Both joined BCHO Partners in 2015, and have begun working as partners since 2019.

Byoungsoo Cho

© 2020 大连理工大学出版社

版权所有·侵权必究

图书在版编目(CIP)数据

都市里的村庄 / 德国里伯斯金建筑事务所等编；司炳月，高松，姜博文译. — 大连：大连理工大学出版社，2020.9
 ISBN 978-7-5685-2657-9

Ⅰ. ①都… Ⅱ. ①德… ②司… ③高… ④姜… Ⅲ. ①城市建筑－建筑设计－世界－现代 Ⅳ. ①TU984

中国版本图书馆CIP数据核字(2020)第155958号

出版发行：大连理工大学出版社
　　　　　（地址：大连市软件园路80号　邮编：116023）
印　　刷：上海锦良印刷厂有限公司
幅面尺寸：225mm×300mm
印　　张：14.75
出版时间：2020年9月第1版
印刷时间：2020年9月第1次印刷
出 版 人：金英伟
统　　筹：房　磊
责任编辑：张昕焱
封面设计：王志峰
责任校对：杨　丹
书　　号：978-7-5685-2657-9
定　　价：298.00元

发　　行：0411-84708842
传　　真：0411-84701466
E-mail：12282980@qq.com
URL：http://dutp.dlut.edu.cn

本书如有印装质量问题，请与我社发行部联系更换。

墙体设计
ISBN：978-7-5611-6353-5
定价：150.00元

新公共空间与私人住宅
ISBN：978-7-5611-6354-2
定价：150.00元

住宅设计
ISBN：978-7-5611-6352-8
定价：150.00元

文化与公共建筑
ISBN：978-7-5611-6746-5
定价：160.00元

城市扩建的四种手法
ISBN：978-7-5611-6776-2
定价：180.00元

复杂性与装饰风格的回归
ISBN：978-7-5611-6828-8
定价：180.00元

内在丰富性建筑
ISBN：978-7-5611-7444-9
定价：228.00元

建筑谱系传承
ISBN：978-7-5611-7461-6
定价：228.00元

伴绿而生的建筑
ISBN：978-7-5611-7548-4
定价：228.00元

微工作·微空间
ISBN：978-7-5611-8255-0
定价：228.00元

居住的流变
ISBN：978-7-5611-8328-1
定价：228.00元

本土现代化
ISBN：978-7-5611-8380-9
定价：228.00元

都市与社区
ISBN：978-7-5611-9365-5
定价：228.00元

木建筑再生
ISBN：978-7-5611-9366-2
定价：228.00元

休闲小筑
ISBN：978-7-5611-9452-2
定价：228.00元

景观与建筑
ISBN：978-7-5611-9884-1
定价：228.00元

地域文脉与大学建筑
ISBN：978-7-5611-9885-8
定价：228.00元

办公室景观
ISBN：978-7-5685-0134-7
定价：228.00元